Lecture Notes in Physics

Edited by H. Araki, Kyoto, J. Ehlers, München, K. Hepp, Zürich
R. Kippenhahn, München, D. Ruelle, Bures-sur-Yvette
H. A. Weidenmüller, Heidelberg, J. Wess, Karlsruhe and J. Zittartz, Köln

Managing Editor: W. Beiglböck

318

W0038506

Bertrand Mercier

An Introduction to the Numerical Analysis of Spectral Methods

Springer-Verlag
Berlin Heidelberg GmbH

Author

Bertrand Mercier
Aerospatiale, Division Systèmes Stratégiques et Spatiaux
Etablissement des Mureaux
Route de Verneuil, F-78130 Les Mureaux, France

ISBN 978-3-662-13757-4   ISBN 978-3-540-46158-6 (eBook)
DOI 10.1007/978-3-540-46158-6

This work is subject to copyright. All rights are reserved, whether the whole or part of the material
is concerned, specifically the rights of translation, reprinting, re-use of illustrations, recitation,
broadcasting, reproduction on microfilms or in other ways, and storage in data banks. Duplication
of this publication or parts thereof is only permitted under the provisions of the German Copyright
Law of September 9, 1965, in its version of June 24, 1985, and a copyright fee must always be
paid. Violations fall under the prosecution act of the German Copyright Law.

© Springer-Verlag Berlin Heidelberg 1989
Originally published by Springer-Verlag Berlin Heidelberg New York in 1989
Softcover reprint of the hardcover 1st edition 1989

2158/3140-543210 – Printed on acid-free paper

EDITORS' PREFACE

This is a translation of report CEA-N-2278, dated 1981, of the French Atomic Energy Commission, titled **Analyse Numérique des Méthodes Spectrales**. The translation was prepared under the auspices of the Institute for Computer Applications in Science and Engineering (ICASE).

We hope that this book will serve as an elementary introduction to the mathematical aspects of spectral methods. The first part of the monograph is a reasonably complete introduction to the theory of Fourier series while the second part lays some foundations for the theory of polynomial expansion methods, in particular Chebyshev expansions.

No monograph of this size can hope to serve as a comprehensive reference to all aspects of spectral methods. The emphasis here is on proving rigorously some fundamental results related to one-dimensional advection and diffusion equations. No applications of the methods are presented and no revisions have been made to account for results subsequent to 1981. The reader interested in recent theory and applications of spectral methods might wish to consult the book by Canuto et al. [5].

May 1988

Nessan Mac Giolla Mhuiris
Mohammed Yousuff Hussaini

AUTHOR'S PREFACE

These notes were written while I was teaching a course on Spectral Methods at the Université Pierre et Marie Curie, Paris, at the request of Professors P.G. CIARLET and P.A. RAVIART, whom I would like to thank here.

They were originally published in French in 1981 as a C.E.A. report.

Their publication in English would certainly not have been possible without the encouragement of Dr. D. GOTTLIEB, Dr. M.Y. HUSSAINI and Dr. R. VOIGT, and the material support of ICASE.

Special thanks are due to the Editors who have not only performed the translation, but also improved the original manuscript.

The support of the French Commissariat à l'Energie Atomique (C.E.A.), and in particular of Professor R. DAUTRAY, Scientific Director, is also acknowledged.

February 1985   B.MERCIER

CONTENTS

INTRODUCTION

"Spectral methods" is the name given to a numerical approach to the solution of partial differential equations. In this approach the solution to the equation is approximated by a truncated series of special functions which are the eigenfunctions of some differential operator.

Part A of this monograph is devoted to Fourier series, sine series, and cosine series. Sections 1 to 4 are a review of some standard properties of Fourier series approximation. Section 5 is devoted to periodic distributions and their development in Fourier series. In particular, we define there the derivative in the <u>periodic distribution sense</u> and this is used in the definition of periodic Sobolev spaces in section 6 where the approximation properties of the truncation operator P_N are reviewed.

An application of these results is given in section 7 where a Galerkin ("spectral") approximation of the equation

$$\frac{\partial u}{\partial t} + a \frac{\partial u}{\partial x} + \frac{\partial}{\partial x} (au) = 0,$$

with periodic boundary conditions, is considered.

The error analysis for this approximation will be based on L^2 estimates obtained using the skew symmetry of the operator L defined by

$$Lu = a \frac{\partial u}{\partial x} + \frac{\partial}{\partial x} (au).$$

The coefficient $a(x)$ is assumed to be smooth, and we show that the accuracy depends only on the smoothness of the initial data u_0. If u_0 is in C^∞, then the error will decrease faster than N^{-s} for any $s > 0$ (this property is known as "spectral accuracy"). On the contrary, if u_0 is discontinuous, then it is well known (see Gottlieb and Orszag [8]) that the Fourier method leads to some undamped oscillations. However, we show that there is still weak convergence with spectral accuracy. In particular, integral quantities are much more accurately captured than pointwise values. This result shows why <u>smoothing</u> is quite useful in the case of discontinuous data.

Section 8 is devoted to the study of the <u>interpolation</u> operator P_C. If u is continuous, $P_C u$ is the truncated Fourier series which coincides with u at some equally-spaced points θ_j. We show that operator P_C enjoys some useful approximation properties in periodic Sobolev spaces. We also show that $P_C u$ can be evaluated easily from the $u(\theta_j)$ by means of the <u>Fast Fourier Transform</u>.

Turning back to the equation

$$\frac{\partial u}{\partial t} + Lu = 0$$

we carry out, in section 9, an error analysis for the <u>collocation</u> (or "pseudospectral") approximation method discussed in section 8.

We review some facts about time discretization in section 10 where we show that <u>explicit</u> schemes can be used with a time step Δt of order $1/N$.

In section 11 we consider the case where a diffusion term is added to the operator L, i.e.,

$$\frac{\partial u}{\partial t} + Au + Lu = 0$$

where the operator A is a second-order operator.

Finally, section 12 gives a brief analysis of the Fourier approximation of the stationary (elliptic) problem

$$Au = f$$

again with periodic boundary conditions.

In Part B, we try to relax the restriction of periodic boundary conditions. The main tool in this latter half of the monograph is to work with polynomials of degree less than or equal to N.

In section 1, we review the main properties of families of polynomials which are orthogonal with respect to the scalar product $(\cdot,\cdot)_\omega$ defined by

$$(u,v)_\omega = \int_I u(x)\ \overline{v(x)}\ \omega(x)dx$$

where ω is a given weight function and I is usually defined as the interval $(-1,+1)$.

The special case where $\omega(x) = (1 - x^2)^{-1/2}$ corresponds to the Chebyshev polynomials. Using the transformation $x = \cos\theta$, we can map I onto $(0,\pi)$ and the Chebyshev series in x then corresponds to a cosine series in θ.

Choosing as interpolation points $x_j = \cos\theta_j$ where the θ_j are equally spaced, we can then compute the interpolant of u with the Fast Fourier Transform. This is why we put such emphasis on the "Chebyshev weight" $\omega(x) = (1 - x^2)^{-1/2}$.

Section 2 discusses the numerical integration formulae of Gauss, Gauss-Radau, and Gauss-Lobatto types. In section 3 we study the approximation properties of the orthogonal projection operator P_N in the space $L_\omega^2(I)$ where the norm is $\|u\|_\omega \equiv (u,u)_\omega^{1/2}$. To this end, we introduce the weighted Sobolev spaces $H_\omega^m(I)$ containing the functions which along with their derivatives up to the order m are in $L_\omega^2(I)$.

We will show that

$$\|u - P_N u\|_{L_\omega^2(I)} \leq C N^{-m} \|u\|_{H_\omega^m(I)},$$

which is quite similar to what was proved for Fourier series in Part A.

Following Canuto and Quarteroni [4] we derive estimates for $\|u - P_N u\|_{H_\omega^m(I)}$ which show a loss of accuracy compared to the results in Part A. The same kind of analysis is performed for the interpolation operator P_C in section 6.

These results are applied in section 5 to the equation

$$\frac{\partial u}{\partial t} + a(x) \frac{\partial u}{\partial x} = 0, \qquad x \in I, \quad t > 0$$

with homogeneous boundary conditions at $x = \pm 1$.

Let $(x_j)_{1 \leq j \leq N}$ denote a set of N points in the interval I; we define the approximate solution u_N to be a polynomial of degree $\leq N$ such that

$$\left(\frac{\partial u_N}{\partial t} + a \frac{\partial u_N}{\partial x}\right)(x_j) = 0, \qquad 1 \leq j \leq N.$$

When a is everywhere positive at least two sets of collocation points are shown to lead to a stable method.

The first set (Gottlieb's method) is

$$x_j = - \cos \frac{j\pi}{N+1} \ , \quad j = 1, \cdots, N$$

The second set is

$$x_j = - \cos \frac{j\pi}{N+\frac{1}{2}} \ , \quad j = 1, \cdots, N.$$

We will carry out an error analysis for both methods.

Explicit time discretization is considered in section 6. The stability condition is shown to be

$$\Delta t < C N^{-2} \ .$$

In section 7 we show how to use the Fast Fourier Transform to speed up the computations. This is not obvious for the first set of collocation points, but it is possible following the argument in Gottlieb [8].

Finally, section 8 is devoted to the heat equation with variable coefficients.

PART A

THE FOURIER SPECTRAL METHOD

1. Review of Hilbert Bases

Let H be a Hilbert space with an inner product denoted by $(\cdot,\cdot)$. The associated norm $\|\cdot\|$ is defined by

$$\|v\| = (v,v)^{1/2} .$$

Recall that a family $\{W_j \in H\}_{j \in I}$, where I is a set (denumerable or nondenumerable) of indices, is said to be orthonormal if

$$(W_j,W_k) = \delta_{jk} \overset{\text{def}}{=} \begin{cases} 1 & \text{if } j = k \\ 0 & \text{otherwise} \end{cases} .$$

Suppose $u \in H$ is given. We can define, for $j \in I$,

$$\hat{u}_j = (u,W_j) .$$

Let $j_1, \cdots, j_n \in I$ be n given indices and

$$u_n = \sum_{k=1}^{n} \hat{u}_{j_k} W_{j_k} .$$

It is easily verified that u_n is the projection of u on the subspace M_n spanned by $\{W_{j_k}\}$, $1 \le k \le n$.

Consequently, $u - u_n$ is orthogonal to M_n, and thus by Pythagorases theorem

$$\| u \|^2 = \| u_n \|^2 + \| u-u_n \|^2 = \sum_{k=1}^{n} | \hat{u}_{j_k} |^2 + \| u-u_n \|^2 \ .$$

We have for all $\{ j_1, \cdots, j_n \}$, the inequality

$$\sum_{k=1}^{n} | \hat{u}_{j_k} |^2 < \| u \|^2 \ ,$$

and it follows that

(1.1) $$\sum_{j \in I} | \hat{u}_j |^2 < \| u \|^2 .$$ (Bessel's Inequality)

In particular, for a given $u \in H$, only a denumerable subset of coefficients $\hat{u}_j$ can be non-zero. Let $\{ j_k \}_{k \in \mathbb{N}}$ be the indices of the coefficients; it can be shown that the sequence $\{ u_n \}_{n=1}^{\infty}$, where

$$u_n = \sum_{k < N} \hat{u}_{j_k} W_{j_k} ,$$

is Cauchy in H (since $\| u_n - u_m \|^2 = \sum_{m < k < n} | \hat{u}_{j_k} |^2 \to 0$ as $m \to \infty$ with $n > m$, from (1.1)) and thus convergent.

Let $u^{\prime} = \lim_{n \to \infty} u_n$. The element $u^{\prime}$ obviously belongs to the completion M of the subspace spanned by $(W_j)_{j \in I}$. As $u-u^{\prime}$ is orthogonal to M, it is deduced that $u^{\prime}$ is the orthogonal projection of u on M.

We say then that the family $(W_j)_{j \in I}$ is a Hilbert basis of H if $M = H$; in other words if

$$u = \lim_{n \to \infty} u_n ,$$

or equivalently, if

$$\|u\|^2 = \sum_{j \in I} |\hat{u}_j|^2. \qquad \text{(the Parseval Equation)}$$

The Hilbert space H is said to be separable if it admits a denumerable Hilbert basis.

2. Simple Examples of Hilbert Bases

In what follows the Hilbert space may be real or complex.

We denote by B(H) the set of all bounded linear operators of H to itself. (T ε B(H) iff M ε $\mathbb{R}$ such that $\|Tv\| \leq M\|v\|$, for all v ε H). We recall the adjoint T* of T is defined by

$$(Tu,v) = (u,T^*v),$$

and that if T = T*, the operator T is said to be <u>Hermitian</u> (the term <u>self-adjoint</u> is, in principle, reserved for unbounded operators).

We recall the following fundamental theorem of spectral theory (see, e.g. Kato [10]).

Theorem 2.1: <u>Let</u> H <u>be a separable Hilbert space and</u> T <u>a compact</u>[*] <u>Hermitian operator. Then there exists a sequence</u> $(\lambda_n)_{n\epsilon \mathbb{N}}$ <u>and</u> $(W_n)_{n\epsilon \mathbb{N}}$ <u>such that</u>

(i) $\lambda_n \epsilon \mathbb{R}$,

(ii) <u>the family</u> $(W_n)_{n\epsilon \mathbb{N}}$ <u>forms a Hilbert basis in</u> H,

(iii) $TW_n = \lambda_n W_n$ <u>for all</u> $n \epsilon \mathbb{N}$.

<u>Example 1.</u> Let $I =]0,\pi[$ and $H = L^2(I)$, the space of measurable functions f defined almost everywhere on I, with complex values such that

$$\int_I |f(\theta)|^2 d\theta < \infty.$$

The space H is a Hilbert space for the inner product

$$(f,g) = \frac{1}{\pi} \int_0^\pi f(\theta)\overline{g(\theta)}d\theta.$$

where the bar denotes complex conjugation. Let $T: L^2(I) \to L^2(I)$ be defined for $f \epsilon L^2(I)$ by $Tf = u$, where u is the solution of Dirichlet problem.

$$-u'' = f$$
$$u(0) = u(\pi) = 0.$$

---

[*]That is to say, T transforms bounded sets into relatively compact sets.

Following the Lax-Milgram lemma we state that

(2.1)
$$\|Tf\|_1 \leq C\|f\|,$$

for some constant C where $\|\cdot\|_1$ denotes the norm of the Sobolev space $H^1(I)$. As the injection of $H^1(I)$ in $L^2(I)$ is compact, (see e.g., [11]), it follows that T is compact.

The eigenfunctions of T satisfy

$$-(\lambda_n W_n)'' = W_n$$

$$W_n(0) = W_n(\pi) = 0,$$

and we have

$$\lambda_n = \frac{1}{n^2} \quad \text{and} \quad W_n(x) = \sqrt{2} \ \sin nx, \qquad \text{for } n \geqslant 1.$$

From Theorem 2.1, we then infer that all functions $u \in L^2(0,\pi)$ may be written in the form

(2.2)
$$u(\theta) = \sum_{n \geqslant 1} \hat{u}_n \ W_n(\theta),$$

with $W_n(x) = \sqrt{2} \ \sin nx$ and

$$\hat{u}_n = (u, W_n) = \frac{\sqrt{2}}{\pi} \int_0^\pi u(\theta) \ \sin n\theta \ d\theta.$$

Example 2.   We now consider the case where the operator T is defined

for $f \in L^2(I)$ by $Tf = u$, where u is the solution of the following Neumann

problem.

$$-u'' + u = f$$

$$u'(0) = u'(\pi) = 0.$$

Here again, the Lax-Milgram lemma establishs inequality (2.1), and consequent-

ly establishs the compactness of the operator T.

The eigenfunctions of the operator T satisfy

$$-(\lambda_n W_n)'' + \lambda_n W_n = W_n,$$

$$W_n'(0) = W_n'(\pi) = 0.$$

It follows that

$$\lambda_n = \frac{1}{1 + n^2} \qquad \text{for} \quad n > 0,$$

$$W_0(x) = 1$$

$$W_n(x) = \sqrt{2} \cos nx \qquad \text{for} \quad n > 1.$$

From Theorem 2.1, we then deduce that all functions $u \in L^2(0,\pi)$ may be

written in the form

(2.3)
$$u(\theta) = \sum_{n > 0} \hat{u}_n W_n(\theta),$$

where $W_0(\theta) = 1$ and

$$\hat{u}_0 = \frac{1}{\pi} \int_0^\pi u(\theta) d\theta \,,$$

while

$$W_n(\theta) = \sqrt{2} \cos n\theta \,,$$

$$\hat{u}_n = \frac{\sqrt{2}}{\pi} \int_0^\pi u(\theta) \cos n\theta \, d\theta \,,$$

for $n > 1$.

Remark 2.1:   Relations (2.2) and (2.3) are termed the expansion of a function u in a sine series or in a cosine series respectively.

If we truncate these expansions at order N, we obtain two different approximations to the function u.   The first (the sine series) gives an approximation of u by a series of functions vanishing at the boundaries which may not converge uniformly to u, if u does not also vanish at the boundaries.   The second (the cosine series) gives an approximation of u by a sum of functions whose first derivative vanishes at the boundaries, and whose derivative may not converge uniformly, to the derivative u´ of u.

We will see that expansions in terms of Fourier series (which are comprised of both of the two preceding expansions) will give us, in general, more satisfactory results.

3. Fourier Series in $L^2(-\pi,\pi)$

We consider now the complex Hilbert space $L^2(-\pi,\pi)$ with a scalar product defined by

$$(f,g) = \frac{1}{2\pi} \int_{-\pi}^{\pi} f(\theta)\overline{g(\theta)}d\theta .$$

We consider also the set $(W_n)_{n \in \mathbb{Z}}$ of trigonometric functions defined by

$$W_n(\theta) = e^{in\theta}.$$

Theorem 3.1: <u>The set $(W_n)_{n \in \mathbb{Z}}$ is a Hilbert basis.</u>

<u>Proof</u>: Any function $u \in L^2(-\pi,\pi)$ is a sum of an even function u_e and an odd function u_o defined by:

$$u_e(x) = \tfrac{1}{2}\left[u(x) + u(-x)\right]$$

$$u_o(x) = \tfrac{1}{2}\left[u(x) - u(-x)\right].$$

From the preceding sections, it follows that for $x \in \,]0,\pi[$, we can expand

$$(3.1) \qquad\qquad u_o(x) = \sum_{n \geq 1} a_n \sin nx$$

$$(3.2) \qquad\qquad u_e(x) = b_0 + \sum_{n \geq 1} b_n \cos nx$$

where

(3.3)
$$a_n = \frac{2}{\pi} \int_0^\pi u_o(\theta) \sin n\theta \, d\theta,$$

(3.4)
$$b_n = \frac{2}{\pi} \int_0^\pi u_e(\theta) \cos n\theta \, d\theta \qquad \text{for } n \geqslant 1,$$

and

$$b_0 = \frac{1}{\pi} \int_0^\pi u_e(\theta) d\theta.$$

For odd or even functions, it can be seen that the relations (3.1) and (3.2) are still valid for $x \, \varepsilon \,]-\pi,\pi[$.

As

$$\cos nx = \tfrac{1}{2}(e^{inx} + e^{-inx})$$

and as

$$\sin nx = \frac{1}{2i}(e^{inx} - e^{-inx}),$$

it can be shown that

$$u(x) = u_o(x) + u_e(x) = b_0 + \sum_{n \geqslant 1} \left[\frac{b_n}{2}(e^{inx} + e^{-inx}) - i\,\frac{a_n}{2}(e^{inx} - e^{-inx}) \right];$$

i.e.,

$$u(x) = \sum_{n \varepsilon \, \mathbb{Z}} \hat{u}_n \, e^{inx}$$

where

$$\hat{u}_0 = b_0$$

and

$$\hat{u}_n = \tfrac{1}{2}(b_n - ia_n)$$

$$\hat{u}_{-n} = \tfrac{1}{2} \left(b_n + ia_n \right) \qquad \text{for} \quad n \geqslant 1.$$

Finally, note that

$$a_n = \frac{1}{\pi} \int_{-\pi}^{\pi} u_o(\theta)\sin n\theta \; d\theta = \frac{1}{\pi} \int_{-\pi}^{\pi} u(\theta)\sin n\theta \; d\theta,$$

$$b_n = \frac{1}{\pi} \int_{-\pi}^{\pi} u_e(\theta)\cos n\theta \; d\theta = \frac{1}{\pi} \int_{-\pi}^{\pi} u(\theta)\cos n\theta \; d\theta,$$

consequently,

$$\hat{u}_n = \frac{1}{2\pi} \int_{-\pi}^{\pi} u(\theta)e^{-in\theta} \; d\theta = (u,W_n).$$

As $(W_n)_{n \in \mathbb{Z}}$ is a complete orthonormal set, it is a Hilbert basis.

Q.E.D.

Corollary 3.1: Let S_N be the subspace of $H \overset{\text{def}}{=} L^2(-\pi,\pi)$ spanned by the functions (e^{inx}), $|n| \leqslant N$ (and of dimension $2N+1$); then the operator $P_N : H \to S_N$ defined by

$$(P_N u)(x) = \sum_{|n| \leqslant N} \hat{u}_n e^{inx}, \qquad \text{for} \quad u \in H,$$

where $\hat{u}_n$ is defined by (3.5), coincides with the orthogonal projection on S_N and satisfies:

$$\| u - P_N u \| \to 0 \quad \text{as} \quad N \to \infty.$$

Remark 3.1:

(1)   For a given $u \in L^2(-\pi,\pi)$, the function $u_N = P_N u$ constitutes an approximation of u in $L^2(-\pi,\pi)$.

The function u_N being periodic, may not converge uniformly to u unless u is periodic and of period 2π.

(2)   Recall that L^2 convergence does not imply convergence almost everywhere; there is no reason for $P_N u$ to converge almost everywhere to u. However, this difficult result is true (see [6]).

The coefficients $\hat{u}_n$ defined by (3.5) are called the <u>Fourier coeffi-</u><u>cients</u> of the function u.

Remark 3.2:   The relationship between the Fourier series and the Fourier transform:

(1)   Suppose u is a periodic function with period 2π; we set

$$f(x) = \begin{cases} u(x) & \text{if } x \in I \equiv (-\pi,\pi) \\ 0 & \text{otherwise} \end{cases},$$

then the Fourier transform $\hat{f}$ of f satisfies

$$\hat{f}(w) \overset{\text{def}}{=} \frac{1}{\sqrt{2\pi}} \int_{\mathbb{R}} e^{-iwx} f(x)dx = \frac{1}{\sqrt{2\pi}} \int_I e^{-iwx} u(x)dx.$$

Therefore

$$\hat{f}(k) = \sqrt{2\pi}\, \hat{u}_k.$$

In other words, the Fourier transform of u (vanishing outside the interval $(-\pi,\pi)$) takes the values $\hat{u}_k \sqrt{2\pi}$ at the points $k \in \mathbb{Z}$.

(2) The Fourier transform of u is a sum of distributions whose weights at the points $k \in \mathbb{Z}$ are precisely the Fourier coefficients of u multiplied by $\sqrt{2\pi}$.

In effect, if

$$\hat{f}(w) = \frac{1}{\sqrt{2\pi}} \int_{\mathbb{R}} f(x) e^{-iwx} \, dx,$$

then

$$f(x) = \frac{1}{\sqrt{2\pi}} \int_{\mathbb{R}} \hat{f}(w) e^{iwx} \, dw.$$

Introducing the Dirac measure δ_{w_0},

$$\hat{f} = \delta_{w_0} \quad \Longrightarrow \quad f(x) = \frac{1}{\sqrt{2\pi}} e^{iw_0 x};$$

therefore,

$$u(x) = \sum_{n \in \mathbb{Z}} \hat{u}_n e^{inx} \quad \Longrightarrow \quad \hat{u}(w) = \sqrt{2\pi} \sum_{n \in \mathbb{Z}} \hat{u}_n \delta_n(w).$$

Remark 3.3: We have used a theorem in spectral theory (Theorem 2.1) to prove the completeness of the Fourier basis $(e^{inx})_{n \in \mathbb{Z}}$ in $L^2(-\pi, \pi)$. The reader should be warned that this is not the usual way of proving completeness. We have chosen to do it this way for two reasons: a) the standard method involves quite lengthy proofs, and b) to justify the name "spectral methods" given by Gottlieb and Orszag [9].

4. The Uniform Convergence of the Fourier Series

Let us observe in the first place that if $I = [-\pi,\pi]$ and $u \in L^2(I)$, then the Fourier coefficients of u are always less than the average of $|u|$ in I

$$(4.1) \qquad |\hat{u}_n| < M(u) \stackrel{def}{=} \frac{1}{2\pi} \int_{-\pi}^{\pi} |u(x)| dx.$$

Moreover, if u is continuous and periodic, with period 2π, and differentiable, then, setting $v = u'$ we have

$$\hat{u}_n = \frac{\hat{v}_n}{in} \, ,$$

(in effect, on integration by parts, we have

$$\hat{u}_n = \frac{1}{2\pi} \int_{-\pi}^{\pi} u(x) e^{-inx} dx = \frac{1}{2\pi} \left[u(x) \frac{e^{-inx}}{-in} \right]_{-\pi}^{+\pi} - \frac{1}{2\pi} \int_{-\pi}^{\pi} u'(x) \frac{e^{-inx}}{-in} dx).$$

More generally, if u is α times differentiable, and has continuous and periodic derivatives up to order $\alpha-1$, we have

$$(4.2) \qquad \hat{u}_n = \frac{\hat{v}_n}{(in)^\alpha} \, ,$$

where $\hat{v}_n$ are the Fourier coefficients of $v = u^{(\alpha)}$. In particular (see (4.1)) we have

$$(4.3) \qquad |\hat{u}_n| < \frac{M(u^{(\alpha)})}{|n|^\alpha} \, .$$

Thus, the more regular a function is, the more rapidly its Fourier coefficients tend to zero as $|n| \to \infty$.

Proposition 4.1: <u>If</u> u <u>is twice continuously differentiable and its</u> <u>first derivative is continuous and periodic with period</u> 2π, <u>then its Fourier</u> <u>series</u> $u_N = P_N u$ <u>converges uniformly to</u> u.

<u>Proof</u>: According to (4.3) we have

$$|\hat{u}_n| < \frac{M_2}{|n|^2}, \qquad \text{for } n \neq 0,$$

where $M_2 = M(u'')$.

The series of moduli (the absolute series)

$$\sum_{n \in \mathbb{Z}} |\hat{u}_n e^{inx}|$$

is less, (independently of x) than the convergent series of positive numbers

$$\hat{u}_0 + \sum_{n \neq 0} \frac{M_2}{n^2}.$$

This proves that the Fourier series of u converges absolutely and uniformly to a continuous function W. Thus converges also in $L^2(I)$ to W since I is bounded. Therefore $W = u$ from Corollary 3.1.

Q. E. D.

5. The Fourier Series of a Distribution

Suppose $I = [-\pi, \pi]$. Let us define $C_p^\infty(I)$ to be the space of functions which are along with all their derivatives, continuous and periodic with period 2π.

From (4.3), we see that functions in $C_p^\infty(I)$ have Fourier coefficients which <u>decrease rapidly</u>; if $\phi \in C_p^\infty(I)$, then, for all $\alpha > 0$, there exists a positive constant C_α such that

$$(5.1) \qquad\qquad |\hat{\phi}_n| < \frac{C_\alpha}{|n|^\alpha} \; .$$

In other words, if $\phi \in C_p^\infty(I)$ then

$$(5.2) \qquad\qquad \text{for all } \beta > 0, \qquad \lim_{|n| \to \infty} |\hat{\phi}_n| \; |n|^\beta \to 0 \; ,$$

(apply (5.1) with $\alpha = \beta + 1$).

Let us call $D_p'(I)$ the dual space of $C_p^\infty(I)$. This is the space of periodic distributions with period 2π.

We will denote by $\langle \cdot, \cdot \rangle$ the duality between $C_p^\infty(I)$ and $D_p'(I)$; if $f \in L^2(I)$ and $\phi \in C_p^\infty(I)$, we have

$$\langle f, \phi \rangle = (f, \phi),$$

where $(\cdot, \cdot)$ is the scalar product of $L^2(I)$ defined previously.

If $f \in D_p'(I)$, we will <u>define</u> the Fourier coefficients $(\hat{f}_n)_{n \in \mathbb{Z}}$ by:

$$\hat{f}_n = \langle f, W_n \rangle \; ,$$

where $W_n(x) = e^{inx}$, (note that $W_n \in C_p^\infty(I)$).

We have for $\phi \in C_p^\infty(I)$

$$\langle f,\phi \rangle = \langle f, \sum_{n \in \mathbb{Z}} \hat{\phi}_n W_n \rangle = \sum_{n \in \mathbb{Z}} \langle f, W_n \rangle \overline{\hat{\phi}}_n ,$$

which implies that

$$\langle f,\phi \rangle = \sum_n \hat{f}_n \overline{\hat{\phi}}_n .$$

Therefore, the series on the right-hand side (of the last equation) should converge for all $\phi \in C_p^\infty(I)$. As the condition of rapid decrease (5.2) holds for functions in $C_p^\infty(I)$, we see that $f \in D_p^-(I)$ iff the sequence of its Fourier coefficients increases slowly, that is:

(5.3) $f \in D_p^-(I)$ iff there exists $k > 0$ such that $\lim\limits_{|n| \to \infty} \dfrac{\hat{f}_n}{(1+n^2)^k} = 0.$

The reciprocal is also true, (cf. Schwartz, [16], p. 225) and results from the fact that any periodic distribution can be represented as a finite sum of the derivatives of continuous functions.

We can now define the derivative in the periodic distribution sense by:

(5.4) $\langle f^{(\alpha)},\phi \rangle \overset{\text{def}}{=} (-1)^\alpha \langle f,\phi^{(\alpha)} \rangle$, for all $\phi \in C_p^\infty(I)$.

The derivative of order α of f is then by definition a periodic distribution

$$g = f^{(\alpha)} \in D_p^-(I).$$

We show that

(5.5)
$$\hat{g}_n = (in)^\alpha \, \hat{f}_n.$$

This results from (4.2) if we write $u = \phi$ and $v = \phi^{(\alpha)}$, for then

$$\langle f^{(\alpha)}, \phi \rangle = (-1)^\alpha \sum_{n \in \mathbb{Z}} \hat{f}_n \, \overline{\hat{v}_n}$$

$$= (-1)^\alpha \sum_{n \in \mathbb{Z}} \hat{f}_n \, \overline{(in)^\alpha \, \hat{\phi}_n}$$

$$= \sum_{n \in \mathbb{Z}} (in)^\alpha \, \hat{f}_n \, \overline{\hat{\phi}_n},$$

which yields the result.

Remark 5.1: The derivative in the periodic distribution sense of a function f which is regular but nonperiodic (i.e., $f(\pi) \neq f(-\pi)$) will introduce a Dirac mass (concentrated mass) at the point π (or at $-\pi$ which coincides with π modulo 2π) as it will for a function discontinuous at a point. The derivative in the sense of $D'_p(I)$ does not coincide with the derivative in the sense of $D'(I)$ (for which relation (4.2) is false).

We will now study the convergence in $D'_p(I)$ of the Fourier series for f:

$$\sum_{n \in \mathbb{Z}} \hat{f}_n \, W_n.$$

(where $W_n(x) = e^{inx}$).

First, we are interested in the case where f is the Dirac distribution δ.

Proposition 5.1: <u>The Fourier series for</u> δ <u>converges in</u> $D_p^-(I)$.

<u>Proof</u>: By definition of Dirac distribution, we have

$$\langle \delta, \phi \rangle = \overline{\phi}(0), \qquad \text{for all} \quad \phi \; \varepsilon \; C_p^\infty(I)$$

therefore $\hat{\delta}_n = 1$, for all $n \; \varepsilon \; \mathbb{Z}$.

We get

$$\delta_N = \sum_{|n| < N} W_n,$$

(δ_N is the Fourier series for δ truncated to the order N). We will show that $\delta_N \to \delta$ in $D_p^-(I)$ when $N \to \infty$.

Suppose $\phi \; \varepsilon \; C_p^\infty(I)$. We have

$$\langle \delta_N, \phi \rangle = \langle \sum_{|n| < N} W_n, \phi \rangle = \sum_{|n| < N} \langle W_n, \phi \rangle = \sum_{|n| > N} \overline{\hat{\phi}_n} = \overline{P_N \phi(0)} \; .$$

Now, we know (see Proposition 4.1) that $P_N \phi$ (Fourier series for ϕ truncated to order N) converges uniformly to ϕ, therefore:

$$\lim_{N \to \infty} \langle \delta_n, \phi \rangle = \overline{\phi}(0) = \langle \delta, \phi \rangle$$

and the result follows.   Q.E.D.

Theorem 5.1: <u>The Fourier series of distribution</u> $f \; \varepsilon \; D_p^-(I)$ <u>converges to</u> f <u>in</u> $D_p^-(I)$.

Proof: For the periodic functions f and g with period 2π, we may define the convolution by:

(5.6)
$$f*g(\theta) = \frac{1}{2\pi} \int_I f(\theta-w)g(w)\,dw.$$

This possesses the usual properties of convolution in $\mathbb{R}$.

Therefore, if $\phi \in C_p^\infty(I)$ and $h = f*g$, we have

$$\langle h,\phi \rangle = \frac{1}{2\pi} \int_I h(\theta')\overline{\phi(\theta')}\,d\theta' = \frac{1}{(2\pi)^2} \int_I \int_I f(\theta'-w)g(w)\overline{\phi(\theta')}\,d\theta'\,dw.$$

Suppose, we set $\theta = \theta' - w$, then

$$\langle h,\phi \rangle = \frac{1}{(2\pi)^2} \int_{I\times I} f(\theta)g(w)\overline{\phi(\theta+w)}\,d\theta\,dw.$$

When $f,g \in D_p'(I)$ are some distributions, we may then generalize the convolution product by setting:

$$\langle f*g,\phi \rangle = \langle f_\theta g_w, \phi(\theta+w) \rangle,$$

where $\langle \cdot,\cdot \rangle$ in the right-hand side denotes the duality between $\left(D_p'(I)\right)^2$ and $\left(C_p^\infty(I)\right)^2$.

Since $f*\delta = f$, we have by continuity of the convolution in the sense of distribution (see e.g., Trêves [18], p. 294)

$$f = f*\delta = \lim_{N\to\infty} f*\delta_N$$

following Proposition 2. Now

$$f * \delta_N = f * \sum_{|n| < N} W_n = \sum_{|n| < N} f * W_n,$$

and

$$(f * W_n)(\theta) = \langle f, \overline{W}_{n,\theta} \rangle,$$

according to (5.6), where

$$W_{n,\theta}(w) \stackrel{\text{def}}{=} W_n(\theta - w) = e^{in\theta} e^{-inw}.$$

Finally

$$(f * W_n)(\theta) = e^{in\theta} \langle f, W_n \rangle = \hat{f}_n W_n(\theta),$$

and so

$$f = \lim_{N \to \infty} \sum_{|n| < N} \hat{f}_n W_n.$$

Q.E.D.

6. Periodic Sobolev Spaces

Let I be the interval $]-\pi, \pi[$. We define the norm $\| \cdot \|$ in the following fashion; for $u \in L^2(I)$, we set

$$\| u \|_r = \left(\sum_{m \in \mathbb{Z}} (1 + m^2)^r |\hat{u}_m|^2 \right)^{1/2},$$

where $\hat{u}_m$ are the Fourier coefficients of u.

We define the space

$$H_p^r(I) = \{u : u^{(\alpha)} \in L^2(I), \alpha = 0, \cdots, r\},$$

where the derivative denoted by the superscript (α) is taken in the periodic distribution sense (see section 5). The space $H_p^r(I)$ is based on the norm

$$|||u|||_r = \left(\sum_{\alpha=0}^{r} C_r^\alpha \|u^{(\alpha)}\|^2\right)^{1/2},$$

where the C_r^α are the usual binomial coefficients and $\|\cdot\|$ denotes the norm of $L^2(I)$ defined in section 3.

We note the following result on denseness.

Lemma 6.1: The space $C_p^\infty(I)$ is dense in $H_p^r(I)$.

Proof: If $u \in H_p^r(I)$, we know that $u_N = P_N u \to u$ in $L^2(I)$ as $N \to \infty$. On the other hand, by the definition of P_N

$$(6.1) \qquad (P_N u)^{(\alpha)} = \sum_{|n| < N} (in)^\alpha \hat{u}_n W_n = P_N u^{(\alpha)},$$

since the coefficients of $u^{(\alpha)}$ are $(in)^\alpha \hat{u}_n$ from (5.5). Therefore $u_N^{(\alpha)}$ converges to $u^{(\alpha)}$ in $L^2(I)$.

Finally, $|||u - u_N|||_r \to 0$ as $N \to \infty$, and the result follows, since $P_N u \in C_p^\infty(I)$.

Q.E.D.

The relation (6.1) shows that the operator P_N commutes with the derivative in the periodic distribution sense.

We define then the space

$$\overset{\bullet}{H}{}^r(I) = \left\{ u \in L^2(I): \|u\|_r < +\infty \right\},$$

where $\|\cdot\|_r$ is the norm associated to the scalar product

$$(u,v)_r = \sum_{m \in \mathbb{Z}} (1 + m^2)^r \, \hat{u}_m \, \hat{v}_m.$$

Proposition 6.1: <u>The space</u> $\overset{\bullet}{H}{}^r(I)$ <u>coincides with the space</u> $H_p^r(I)$ and <u>the norm</u> $|||\cdot|||_r$ <u>with the norm</u> $\|\cdot\|_r$.

<u>Proof:</u> Suppose that $u \in \overset{\bullet}{H}{}^r(I)$; it can be deduced that

$$\sum_{m \in \mathbb{Z}} m^{2\alpha} \, |\hat{u}_m|^2 < +\infty, \qquad 0 \leqslant \alpha \leqslant r,$$

Consequently $u^{(\alpha)} \in L^2(I)$ for $0 \leqslant \alpha \leqslant r$.

The converse follows immediately.

Finally, we verify that

$$|||u|||_r^2 = \sum_{\alpha=0}^{r} C_r^\alpha \sum_{m \in \mathbb{Z}} m^{2\alpha} \, |\hat{u}_m|^2 = \sum_{m \in \mathbb{Z}} (1+m^2)^r \, |\hat{u}_m|^2 = \|u_r\|^2.$$

Q.E.D.

The definition of $\overset{\bullet}{H}{}_p^r(I)$ is such that it is sensible for noninteger values of r. The previous result permits the definition of $H_p^r(I)$ to be extended to $r \in \mathbb{R}$.

We are now in a position to give error estimates in the periodic Sobolev spaces.

Theorem 6.1: Let $r, s \in \mathbb{R}$ with $0 \leqslant s \leqslant r$; then we have

$$\| u - P_N u \|_s \leqslant (1+N^2)^{\frac{s-r}{2}} \| u \|_r, \qquad \text{for } u \in H_p^r(I).$$

Proof: We have

$$\| u - P_N u \|_s^2 = \sum_{|m| > N} (1+m^2)^{s-r+r} | \hat{u}_m |^2 \leqslant (1+N^2)^{s-r} \sum_{|m| > N} (1+m^2)^r | \hat{u}_m |^2$$

$$\leqslant (1+N^2)^{s-r} \| u \|_r^2.$$

Q.E.D.

Remark 6.1: The preceding result shows that the more regular u is the better an approximation $P_N u$ is to u. More precisely, if $u \in H_p^r(I)$, we have an error estimate of order $O(N^{-r})$ in norm $L^2(I)$ which clearly improves as r increases.

Lemma 6.2: (Sobolev Inequality). There exists a constant C such that

$$\| u \|_{L^\infty(I)}^2 \leqslant C \| u \|_0 \| u \|_1, \qquad \text{for all } u \in H_p^1(I),$$

and in particular $H_p^1(I) \to L^\infty(I)$.

<u>Proof</u>: Suppose $u \in C_p^\infty(I)$. We know that $\hat{u}_0$ is the average of u over I. From the mean value theorem, there exists $x_0 \in I$ such that $\hat{u}_0 = u(x_0)$. Let $v(x) = u(x) - \hat{u}_0$; we have

$$\tfrac{1}{2}|v(x)|^2 = \int_{x_0}^x v(y)v'(y)dy < \left(\int_{x_0}^x |v(y)|^2 dy\right)^{1/2}\left(\int_{x_0}^x |v'(y)|^2 dy\right)^{1/2} < 2\pi\|v\|\,\|v'\|,$$

$$|u(x)| < |\hat{u}_0| + |v(x)| < |\hat{u}_0| + 2\pi^{1/2}\|u\|^{1/2}\|u'\|^{1/2},$$

because $v' = u'$ and $\|v\| < \|u\|$. Since $|\hat{u}_0| < \|u\|$, we have

$$|u(x)| < C\|u\|_0^{1/2}\|u\|_1^{1/2}.$$

The inequality is then proved for all $u \in C_p^\infty(I)$. Since $C_p^\infty(I)$ is dense in $H_p^1(I)$, it also holds for $u \in H_p^1(I)$.

Q.E.D.

From Lemma 6.2 we immediately obtain an error estimate in $L^\infty(I)$ norm; we have

$$\|u-P_N u\|_{L^\infty(I)}^2 < C(1+N^2)^{-r/2}(1+N^2)^{\frac{1-r}{2}} = C(1+N^2)^{1/2-r},$$

thus

$$\|u-P_N u\|_{L^\infty(I)} = O(N^{1/2-r}),$$

valid for $r > 1$, and uniform convergence for all $u \in H_p^1(I)$. (Note that in this case u is continuous.) This result is stronger than that given in

Proposition 4.1.

　　Remark 6.2:　　If, instead of S_N we consider the space $\widetilde{S}_N$ spanned by the functions (e^{inx}), $-N+1 \leqslant n \leqslant N$, we have some analogous approximation properties for the projection operator $\widetilde{P}_N : L^2 \to \widetilde{S}_N$. (We note that the space $\widetilde{S}_N$ is of dimension $2N$, and the space S_N is of dimension $2N+1$).

7. First-Order Equations — The Galerkin Method

Let L be the first-order operator defined by

$$Lu \equiv a \frac{\partial u}{\partial x} + \frac{\partial (au)}{\partial x} \, ,$$

where $a \in C_p^\infty(I)$ is regular and periodic (real).

We observe first that L is skew symmetric:

$$(Lu,v) = \frac{1}{2\pi} \int_{-\pi}^{\pi} \left(a \frac{\partial u}{\partial x} + \frac{\partial (au)}{\partial x} \right) \overline{v} dx = \frac{-1}{2\pi} \int_{-\pi}^{\pi} \left(u \frac{\partial (a\overline{v})}{\partial x} + au \frac{\partial \overline{v}}{\partial x} \right) dx = -(u,Lv),$$

for $u,v \in D(L) \overset{def}{=} H_p^1(I)$. (Note that we have used the periodicity of a, u, and v in the integration by parts.) We observe that L is a bounded operator of $H_p^k(I)$ in $H_p^{k-1}(I)$.

We consider then the following problem in the space $L^2(I)$. Find $u(t) \in D(L)$ such that

$$\frac{\partial u}{\partial t} + Lu = 0 \qquad\qquad t > 0$$

(7.1)

$$u(x,0) = u_0(x)$$

where $u_0 \in D(L)$ is given.

We have the following existence result:

Theorem 7.1: <u>Let $s > 1$ and $u_0 \in H_p^s(I)$; then the problem (7.1) admits a unique solution $u \in C^0(0,T;H_p^s(I))$. Moreover, there exists a constant C independent of u_0 and t such that:</u>

$$\|u(\cdot,t)\|_s \leq C\|u_0\|_s, \qquad \text{for } t \in [0,T],$$

where T is positive and given.

Proof: The proof of this result is an elegant applicaton of the theory of pseudo-differential operators (see M. Taylor, [17], pp. 62-65). Let us content ourselves with establishing the a priori estimate in H_p^r of the solution assuming it exists.

For that purpose we introduce the operator

$$\Lambda^s : H_p^s(I) \rightarrow L^2(I)$$

defined by

$$u = \sum_{n \in \mathbb{Z}} \hat{u}_n e^{inx} \rightarrow \Lambda^s u \equiv \sum_{n \in \mathbb{Z}} (1+n^2)^{s/2} \hat{u}_n e^{inx}.$$

We note that

$$\|u\|_s = \|\Lambda^s u\|_0;$$

on the other hand, if $s = 2$, we have

$$\Lambda^s u = \left(1 - \frac{d^2}{dx^2}\right)u ,$$

and if s is a multiple of 2, we have

$$\Lambda^s u = \left(1 - \frac{d^2}{dx^2}\right)^{s/2} u.$$

(In the general case where s is real and positive, Λ^s is a pseudo-differential operator of order s.)

If u is a solution of (7.1) we have then, by setting

$$K = [\Lambda^s, L] \equiv \Lambda^s L - L\Lambda^s$$

$$\frac{d}{dt} \|u(t)\|_s^2 = \frac{d}{dt} \|\Lambda^s u(t)\|_0^2 = \left(\Lambda^s \frac{\partial u}{\partial t}, \Lambda^s u\right) + \left(\Lambda^s u, \Lambda^s \frac{\partial u}{\partial t}\right)$$

$$= -(\Lambda^s Lu, \Lambda^s u) - (\Lambda^s u, \Lambda^s Lu)$$

$$= -(L\Lambda^s u, \Lambda^s u) - (Ku, \Lambda^s u) - (\Lambda^s u, L\Lambda^s u) - (\Lambda^s u, Ku)$$

$$= -(Ku, \Lambda^s u) - (\Lambda^s u, Ku),$$

where we have utilized the antisymmetry of L.

Since K is an operator (pseudo-differential in the general case) of order s, it follows that

$$\|Ku\|_0 \leqslant C\|u\|_s,$$

$$\frac{d}{dt} \|u(t)\|_s^2 \leqslant 2\|Ku\|_0 \|\Lambda^s u\|_0 \leqslant 2C\|u\|_s^2.$$

Therefore,

$$\|u(t)\|_s^2 \leqslant e^{2Ct} \|u_0\|_s,$$

and the result follows.

Let us verify in the case   s = 2   that   K   is truely an operator of order   s   (and not of order   s+1); in this case

$$Ku = \left(1 - \frac{d^2}{dx^2}\right)(Lu) - L(u-u'')$$

whence by setting $Lu = bu' + C$ with $b = 2a$ and $C = a'$, we get,

$$Ku = bu' + C - (bu'+C)'' - b(u-u'')' - C(u-u'')$$

$$= bu' + C - (b''u'+2b'u''+bu''') - C'' - b(u'-u''') - C(u-u''),$$

and we see that the terms of third order disappear.   Q. E. D.

We carry out now a __spatial semi-discretization__ of the problem (7.1) by looking for an approximate solution $u_N(t) \equiv u_N(\cdot,t)$ in the space S_N spanned by the functions $(e^{inx})_{|n|\leqslant N}$ [1].

The approximate problem is therefore the following.

Find u_N such that

$$\left(\frac{\partial u_N}{\partial t} + Lu_N, v_N\right) = 0, \qquad \text{for all}\quad v_N \in S_N, \quad t > 0$$

(7.2)

$$\left(u_N(0) - u_0, v_N\right) = 0, \qquad \text{for all}\quad v_N \in S_N.$$

[1] See Corollary 3.1.

Let $L_N = P_N L$, where $P_N : L^2(I) \to S_N$ is the projection on S_N, we may equivalently write (7.2) in the form

$$\frac{\partial u_N}{\partial t} + L_N u_N = 0$$

(7.3)

$$u_N(0) = P_N u_0.$$

Note that L_N is <u>also antisymmetric</u>

$$(L_N u_N, v_N) = (Lu_N, v_N) = - (u_N, Lv_N) = - (u_N, L_N v_N)$$

for u_N, $v_N \in S_N$. In particular, with $u_N = v_N$

(7.4) $$Re(L_N v_N, v_N) = 0.$$

Since S_N is of finite dimension, the equation (7.3) is in fact a differential system with a solution $u_N \in C^0(0,T;S_N)$ [in other words the operator L_N generates a semi-group of class C^0].

We establish the convergence result in the following

Theorem 7.2: <u>If</u> $u_0 \in H_P^{s+1}(I)$, <u>and</u> u <u>is the solution of the equation</u> (7.1) <u>then there exists a constant</u> C_1 <u>independent of</u> u_0 <u>and</u> t <u>such that</u>

$$\| u(t) - u_N(t) \|_0 \leqslant C_1 (1+N^2)^{-s/2} \| u_0 \|_{s+1}, \qquad \text{for } t \in [0,T].$$

Proof: We set $\tilde{u}_N(t) = P_N u(t)$ and note that

$$\left(\frac{\partial u}{\partial t}\right)_n = \left(\frac{\partial u}{\partial t}, W_n\right) = \frac{d}{dt}(u, W_n) = \frac{d}{dt}\hat{u}_n$$

therefore

$$P_N \frac{\partial u}{\partial t} = \frac{\partial}{\partial t} P_N u = \frac{\partial}{\partial t}\tilde{u}_N \ .$$

Consequently, (see (7.1))

$$\frac{\partial \tilde{u}_N}{\partial t} + L_N u = 0$$

therefore

$$\frac{\partial \tilde{u}_N}{\partial t} + L_N \tilde{u}_N = L_N(\tilde{u}_N - u).$$

Subtracting from (7.3) and setting $W_N = u_N - \tilde{u}_N$, we obtain

$$\frac{\partial W_N}{\partial t} + L_N W_N = L_N(u - \tilde{u}_N).$$

Taking the scalar product with W_N, we obtain

$$\left(\frac{\partial W_N}{\partial t}, W_N\right) + (L_N W_N, W_N) = (L_N(u - \tilde{u}_N), W_N).$$

whence (taking the real part, and applying (7.4)):

$$\frac{1}{2}\frac{d}{dt}\|W_N(t)\|_0^2 = Re(L_N(u - \tilde{u}_N), W_N) < \|L_N(u - \tilde{u}_N)\|_0 \|W_N\|_0.$$

Utilizing the identity

$$\frac{1}{2}\frac{d}{dt}\|W_N(t)\|_0^2 = \|W_N(t)\|_0 \frac{d}{dt}\|W_N(t)\|_0 \ ,$$

we obtain

(7.5) $\qquad \frac{d}{dt} \| W_N(t) \|_0 < \| L_N(u-\tilde{u}_N) \|_0 < \| L(u-\tilde{u}_N) \|_0 < C_2 \| u-\tilde{u}_N \|_1$

where the constant C_2 is the constant of continuity for the mapping

$$L : H_P^1(I) \to L^2(I).$$

Since

$$\| (u-\tilde{u}_N)(t) \|_1 < (1+N^2)^{-s/2} \| u(t) \|_{s+1} < C(1+N^2)^{-s/2} \| u_0 \|_{s+1} \, ,$$

according to the Theorems 6.1 and 7.1, we deduce from (7.5)

$$\| W_N(t) \|_0 < CC_2(1+N^2)^{-s/2} t \| u_0 \|_{s+1};$$

as

$$\| (u-u_N)(t) \|_0 < \| (u-\tilde{u}_N)(t) \|_0 + \| W_N(t) \|_0;$$

we have obtained the desired result and an evaluation of the constant C.

$\qquad\qquad\qquad\qquad\qquad\qquad\qquad\qquad\qquad\qquad\qquad$ Q.E.D.

Remark 7.1:

(1) If $u_0 \in H_P^{s+1}(I)$, we have therefore an error estimate of $O(N^{-s})$ in the norm $\| \cdot \|_0$, with a constant which increases linearly with t.

The method is thus of <u>infinite order</u> in the sense that the accuracy of the method is only limited by the regularity of the initial data (and the coefficients). If this is in $C_p^\infty(I)$, the error decreases to zero as $N \to \infty$ faster than N^{-s} for all $s > 0$. This property is called "spectral accuracy."

This shows that the spectral methods will be superior to all the finite element or finite difference methods from the point of view of accuracy when one is dealing with regular solutions.

(2) We may replace the space S_N (of dimension 2N+1) by the space $\tilde{S}_N$ (of dimension 2N) introduced by Remark 6.2, with exactly the same results.

Remark 7.2:   Estimate in the norm of Sobolev spaces of negative indices.

Let Φ be a given function sufficiently regular; we will show that the quantity $(\Phi, u_N(t) - u(t))$ converges "sufficiently rapidly" to zero as $N \to \infty$ even if $u(t)$ is not regular.

For that purpose, we introduce the solution W of the adjoint problem

$$\frac{\partial W}{\partial t} + L^* W = 0, \qquad t > 0$$

$$W(0) = \Phi$$

where L^* ($= -L$) is the adjoint of L.

Let $W_N(t) \in S_N$ be the solution of the approximate adjoint problem

$$\left(\frac{\partial W_N}{\partial t} + L^* W_N, v_N\right) = 0, \qquad \text{for all } v_N \in S_N$$

$$\left(W_N(0) - \Phi, v_N\right) = 0, \qquad \text{for all } v_N \in S_N.$$

According to the Theorem 7.2, if $\Phi \in H_P^{s+1}(I)$,

$$\|W(t) - W_N(t)\|_0 \leq C N^{-s} \|\Phi\|_{s+1}$$

for $t \leq T$.

Using the relation

$$(7.6) \qquad \bigl(\Phi, u_N(t) - u(t)\bigr) = \bigl(W_N(t) - W(t), u_0\bigr),$$

(which we will establish shortly) we deduce the upper bound sought

$$(7.7) \qquad \bigl(\Phi, u_N(t) - u(t)\bigr) \leq C N^{-s} \|\Phi\|_{s+1} \|u_0\|_0.$$

Noting that

$$\|u_N(t) - u(t)\|_{-\sigma} \equiv \sup_{\Phi \in H_P^{\sigma}(I)} \frac{\bigl(\Phi, u_N(t) - u(t)\bigr)}{\|\Phi\|_{\sigma}},$$

we may interpret (7.7) as an error estimate in the Sobolev space of negative indices.

In the extreme case where u_0 is discontinuous, we observe then on account of the Gibbs phenomenon an oscillation in the approximate solution in the vicinity of the discontinuity, but the oscillations annul themselves "in the mean," according to the relation (7.7) (since the second member of (7.7) converges to zero as Φ is regular).

This explains intuitively the success of the Fourier method with smoothing, consisting of smoothing the initial solution u_0 (see Majda-McDonough-Osher, [12]). By that we mean the following; let ρ be a positive regular function with a compact support such that:

$$\int \rho(x)dx = 1.$$

We set

$$\rho_\varepsilon(x) = \rho\left(\frac{x}{\varepsilon}\right)$$

and

$$u_\varepsilon(t) = \rho_\varepsilon * u(t)$$

$$u_{\varepsilon N} = \rho_\varepsilon * u_N.$$

We know that $u_\varepsilon(t) \to u(t)$ when $\varepsilon \to 0$, since by definition

$$u_\varepsilon(x) = \int \rho_\varepsilon(x-y)u(y)dy = (u, \rho_{\varepsilon x})$$

(where we have set $\rho_{\varepsilon x}(y) \overset{\text{def}}{=} \rho_\varepsilon(x-y)$). We have

$$\left|(u_\varepsilon - u_{\varepsilon N})(x)\right| = \left|(\rho_{\varepsilon x}, u_N - u)\right| < CN^{-(s-1)} \|\rho_{\varepsilon x}\|_s \|u_0\|_0$$

as

$$\|\rho_{\varepsilon x}\|_s = \|\rho_\varepsilon\|_s.$$

We deduce that if ρ is very regular, for all $\varepsilon > 0$, there exists a constant $C(s,\varepsilon)$ such that

$$\left| (u_\epsilon - u_{\epsilon N})(x) \right| < C(s,\epsilon) N^{-s} \ ,$$

Therefore, there is uniform convergence of the regularized u_N to the regularized u, which has an "infinite rate of convergence."

Proof of (7.6):   We have by definition

$$\left(\Phi, \ u_N(t) - u(t) \right) = \left(W_N(0), u_N(t) \right) - \left(W(0), u(t) \right).$$

Now,

$$\left(W_N(0), u_N(t) \right) = \left(W_N(t), \ u_N(0) \right) + \int_0^t \frac{d}{ds} \left(\tilde{W}_N(s), u_N(s) \right) ds$$

where have set

$$\tilde{W}_N(s) = W_N(t-s).$$

Noting that $\tilde{W}_N'(s) = -W_N'(t-s)$ we have

$$\int_0^t \frac{d}{ds} \left(\tilde{W}_N(s), u(s) \right) ds = \int_0^t \left((W_N(t-s), u_N'(s)) - (W_N'(t-s), u_N(s) \right) ds$$

$$= \int_0^t \left(W_N(t-s), -Lu_N(s) \right) - \left(-L^* W_N(t-s), u_N(s) \right) ds$$

$$= 0,$$

which yields

$$\left(W_N(0), u_N(t) \right) = \left(W_N(t), u_N(0) \right).$$

It can be shown that

$$\big(W(0),u(t)\big) = \big(W(t),u(0)\big),$$

whence

$$\big(\Phi,u_N(t)-u(t)\big) = \big(W_N(t),u_N(0)\big) - (W(t),u(0))$$

and result (7.6) follows.   Q.E.D.

8. Lagrange Interpolation in S_N; The Discrete Fourier Transform

In practice, if $u \in C_p^0(I)$ is a continuous periodic function on the interval $I = [-\pi,\pi]$, it is not possible to calculate exactly the Fourier coefficients $\hat{u}_N$ of u.

We therefore do not know in general $P_N u$ which is the best approximation of u in S_N (for the norm of $L^2(I)$). However, we will see that it is easy to determine a function $v \in S_N$, called the interpolant of u, which coincides with u at $2N+1$ points $(x_j)_{|j| \leqslant N}$ defined by

$$x_j = jh, \qquad\qquad |j| \leqslant N$$

(8.1) where

$$h = \frac{2\pi}{2N+1}$$

In fact if we set

$$v(x) = \sum_{|k| \leqslant N} a_k e^{ikx},$$

we see that the $2N+1$ coefficients a_k are solutions of the linear system

(8.2)
$$\sum_{|k|<N} e^{ikx_j} a_k = u(x_j), \qquad |j| < N.$$

Now, up to a multiplicative factor $(2N+1)$, the $(2N+1) \times (2N+1)$ matrix of this linear system is <u>unitary</u> (and hence invertible).

In effect (8.2) may be rewritten as

(8.3)
$$\sum_{|k| < N} W^{jk} a_k = u(x_j) \qquad |j| < N$$

where $W = e^{ih} = e^{\frac{2i\pi}{2N+1}}$ is the principal root of order $(2N+1)$ of unity, and we have the identity

(8.4)
$$\frac{1}{2N+1} \sum_{|j|<N} W^{jk} W^{-j\ell} = \delta_{k\ell} = \begin{cases} 1 & \text{if } k = \ell \\ 0 & \text{otherwise} \end{cases},$$

which results from the following lemma (applied with $\omega = W^{k-\ell}$).

Lemma 8.1: <u>Suppose</u> ω <u>is a root of order</u> $2N+1$ <u>of unity; then we have</u>

$$\frac{1}{2N+1} \sum_{|j|<N} \omega^j = \begin{cases} 1 & \underline{\text{if } \omega = 1} \\ 0 & \underline{\text{otherwise}} \end{cases}.$$

Proof: Set $M = 2N+1$ and

$$j = \begin{cases} j & \text{if } 0 < j < N \\ j+M & \text{if } -N < j < 0 \end{cases}.$$

Since $\omega^{j+M} = \omega^j$, we have

$$\frac{1}{2N+1} \sum_{|j| < N} \omega^j = \frac{1}{M} \sum_{j'=0}^{M-1} \omega^{j'} \ .$$

This gives the desired result with the identity

$$(1-\omega) \sum_{j'=0}^{M-1} \omega^{j'} \equiv 1-\omega^M,$$

valid for all $\omega \in \mathbb{C}$.

Q. E. D.

Corollary 8.1: Let $(a_k)_{|k| < N}$ be the Fourier coefficients of the interpolant of u in S_N defined by (8.3), we have:

(8.5)
$$a_k = \frac{1}{2N+1} \sum_{|j| < N} W^{-jk} z_j \ ,$$

where the $(z_j)_{|j| < N}$ are the values of u at x_j given by:

$$z_j = u(x_j) \qquad |j| < N.$$

Definition 8.1: We call **discrete Fourier transform** the mapping

$$(z_j)_{|j| < N} \rightarrow (a_k)_{|k| < N} \ .$$

Remark 8.1: The advantage of the discrete Fourier transform is, that thanks to the existence of the Fast Fourier Transform (see, for example, Auslander-Tolimieri, [1]), the computation of the z_j from the a_k and of the a_k from the z_j can be performed in $O(N \log N)$ operations and not in $O(N^2)$ operations as one would expect when one calculates the product of a full $(2N+1) \times (2N+1)$ matrix by a vector.

In what follows, the mapping which associates with each u its interpolant $v \in S_N$, will be denoted by $P_C : C_P^0(I) \to S_N$. Let $(\cdot,\cdot)_N$ be the sesquilinear form on $C_P^0(I)$ defined by

$$(8.6) \qquad (u,v)_N = \frac{1}{2N+1} \sum_{|j| \leq N} u(x_j) \overline{v(x_j)}.$$

The operator P_C satisfies

$$(P_C u)(x_j) = u(x_j), \qquad |j| \leq N$$

and, in particular

$$(8.7) \qquad (u-P_C u, v_N)_N = 0, \qquad \text{for all} \quad v_N \in S_N.$$

By the definition of P_N

$$(u-P_N u, v_N) = 0, \qquad \text{for all} \quad v_N \in S_N,$$

so we see that in order to obtain P_C, it suffices to replace the scalar product $(\cdot,\cdot)$ by the "discrete scalar product" $(\cdot,\cdot)_N$.

The name "discrete scalar product" may be justified by noting that $(\cdot,\cdot)_N$ and $(\cdot,\cdot)$ coincide on S_N.

$$(8.8) \qquad (u_N,v_N)_N = (u_N,v_N), \qquad \text{for all} \quad u_N,v_N \in S_N.$$

This results from the fact that the <u>numerical integration formula</u>

$$(8.9) \qquad \frac{1}{2\pi} \int_{-\pi}^{\pi} f(x)dx \simeq \frac{1}{2N+1} \sum_{|j| \leq N} f(x_j),$$

is exact for $f \in S_{2N}$.

Indeed, from Lemma 8.1, we have

$$\frac{1}{2N+1} \sum_{|j| \leq N} e^{ikx_j} = \frac{1}{2N+1} \sum_{|j| \leq N} w^{jk} = \begin{cases} 1 & \text{if } k = 0 \ (\text{mod } 2N+1) \\ 0 & \text{otherwise} \end{cases},$$

and thus

$$\frac{1}{2N+1} \sum_{|j| \leq N} e^{ikx_j} = \frac{1}{2\pi} \int_I e^{ikx}dx, \qquad \text{if } |k| < 2N.$$

The Relation Between the Fourier Coefficients of a Function and the Fourier Coefficients of its Interpolant.

Lemma 8.2: <u>Let</u> $u \in C_p^0(I)$ <u>with</u> $(\hat{u}_k)_{k \in \mathbb{Z}}$ <u>as its Fourier coefficients, and</u> $(a_n)_{|n| \leq N}$ <u>the Fourier coefficients of its interpolant</u> $P_C u$ <u>in</u> S_N; <u>we have the relation</u>

$$a_n = \sum_{\ell \in \mathbb{Z}} \hat{u}_{n+\ell M}$$

<u>where</u>

$$M \overset{\text{def}}{=} 2N+1.$$

<u>Proof</u>: Let $(W_n)_{n \in \mathbb{Z}}$ be the basis of $L^2(I)$ defined by

$$W_n(x) = e^{inx}.$$

We have

$$(W_k, W_n)_N = \frac{1}{2N+1} \sum_{|j| < N} W_k(x_j) \overline{W_n(x_j)}$$

$$= \frac{1}{2N+1} \sum_{|j| < N} e^{i(k-n)x_j}$$

$$= \frac{1}{2N+1} \sum_{|j| < N} W^{j(k-n)}.$$

Using Lemma 8.1, applied with $\omega = W^{k-n}$, we may deduce that

$$(8.10) \qquad (W_k, W_n)_N = \begin{cases} 1 & \text{if } k = n \pmod{M} \\ 0 & \text{otherwise} \end{cases}.$$

Since

$$(8.11) \qquad P_C u = \sum_{|n| < N} a_n W_n,$$

we infer from (8.7) and (8.10) that

$$a_n = (P_C u, W_n)_N = (u, W_n)_N$$

$$= \left(\sum_{k \in \mathbb{Z}} \hat{u}_k W_k, W_N \right)_N = \sum_{\ell \in \mathbb{Z}} \hat{u}_{n+\ell M}.$$

Q.E.D.

Remark 8.2: With the preceding notation, we have

(8.12)
$$\hat{u}_k = (u, W_k) = \frac{1}{2\pi} \int_I u\overline{W}_k \, dx$$

(8.13)
$$a_k = (u, W_k)_N = \frac{1}{2N+1} \sum_{|j| \leq N} u(x_j) \overline{W_k(x_j)} \, ,$$

which shows that if we use the numerical integration formula (8.9) to evaluate the integral defining a Fourier coefficient $\hat{u}_k$, we obtain (not $\hat{u}_k$ but) a_k, that is to say the Fourier coefficient of the interpolant of u.

Estimation of $\| u - P_C u \|_0$.

We will establish the following theorem:

Theorem 8.1: Let $r > \frac{1}{2}$ be fixed; then there exists a constant C such that

(8.14)
$$\| u - P_C u \|_0 \leq C \, N^{-r} \| u \|_r, \qquad \text{for all } u \in H_P^r(I).$$

Proof: Noting first that P_C leaves S_N invariant, we have $P_C P_N = P_N$. We may thus write

(8.15)
$$u - P_C u = u - P_N u + P_C(P_N - I)u.$$

. Therefore, by setting $v = (I - P_N)u$

$$\|u - P_C u\|_0 \leqslant \|u - P_N u\|_0 + \|P_C v\|_0.$$

Using Theorem 6.1, it suffices to show that

(8.16)
$$\|P_C v\|_0 \leqslant C N^{-r} \|u\|_r.$$

For this purpose, we note that if the a_k denote the Fourier coefficients of $P_C v$, we have from Lemma 8.12

(8.17)
$$a_k = \sum_{\ell \in \mathbb{Z}} \hat{v}_{k+\ell M},$$

where the $\hat{v}_n$ are the Fourier coefficients of v and satisfy

$$\hat{v}_n = \begin{cases} 0 & \text{if } |n| \leqslant N \\ \hat{u}_n & \text{otherwise} \end{cases}$$

Suppose

$$Y(k) = \{n \in \mathbb{Z} : n = k + \ell M \quad \text{with } \ell \in \mathbb{Z}/\{0\}\},$$

we see that (8.17) may be rewritten

$$a_k = \sum_{n \in Y(k)} \hat{v}_n = \sum_{n \in Y(k)} (1+n^2)^{-r/2} \, \hat{v}_n (1+n^2)^{r/2} \, .$$

Using the Cauchy-Schwarz inequality, we have then that

(8.18)
$$|a_k| \leqslant \Big(\sum_{n \in Y(k)} (1+n^2)^{-r} \Big)^{1/2} \Big(\sum_{n \in Y(k)} (1+n^2)^r \, |\hat{v}_n|^2 \Big)^{1/2} \, .$$

Now,

(8.19)
$$\sum_{n \in Y(k)} (1+n^2)^{-r} < CN^{-2r}.$$

In fact,

$$\sum_{n \in Y(k)} (1+n^2)^{-r} = N^{-2r} \sum_{n \in Y(k)} \frac{1}{(\frac{1}{N^2} + \frac{n^2}{N^2})^r} = N^{-2r} \sum_{\ell \in \mathbb{Z}/\{0\}} \frac{1}{b_\ell^r},$$

where

$$b_\ell \equiv \frac{1}{N^2} + \left(\frac{k+\ell M}{N}\right)^2.$$

Now,

$$b_\ell \geqslant \ell^2,$$

(since $M = 2N+1$ and $|k| < N$) so the series

$$\sum_{\ell > 0} b_\ell^{-r} < \sum_{\ell > 0} \ell^{-2r} \overset{\text{def}}{=} \sigma(r) < +\infty.$$

We deduce (8.19) with $C = 2\sigma(r)$. Returning to (8.18) we see that

$$|a_k| < CN^{-2r} \sum_{n \in Y(k)} (1+n^2)^r |\hat{v}_n|^2.$$

Thus

$$\|P_C v\|_0^2 = \sum_{|k|<N} |a_k|^2 < CN^{-2r} \sum_{|k|<N} \sum_{n\varepsilon Y(k)} (1+n^2)^r |\hat{v}_n|^2$$

$$< CN^{-2r} \sum_{n\varepsilon \mathbb{Z}} |\hat{v}_n|^2 (1+n^2)^r < CN^{-2r} \sum_{n\varepsilon \mathbb{Z}} |\hat{u}_n|^2 (1+n^2)^r$$

$$< CN^{-2r} \|u\|_r^2 ,$$

and (8.16) follows.   Q.E.D.

Remark 8.3: From the formula (8.18) we see that $|a_k|$ is bounded since $r > 1/2$. Thus P_C is defined when $u \varepsilon H_P^r(I)$ for $r > 1/2$. This is consistent because of the results of injection of the Sobolev space $H_P^r(I)$ into $C_P^0(I)$ when $r > 1/2$.

Remark 8.4: We have defined the discrete Fourier transforms for only an odd number $(2N+1)$ of points. We can define the discrete transform for an even number of points constructing an interpolant in the space $\tilde{S}_N$ (introduced in Remark 6.2) and which is of dimension $2N$. To do this we choose

$$\tilde{x}_j = j\tilde{h}, \qquad -N + 1 < j < N$$
$$\tilde{h} = \frac{\pi}{N} .$$

The analogue of (8.2) is then

$$\sum_{k=-N+1}^{N} e^{ik\tilde{x}_j} a_k = u(\tilde{x}_j) \qquad -N + 1 < j < N ,$$

whose matrix, multiplied by a factor $2N$, is unitary. The analogue of Lemma

8.1 is easy to establish, and is given as follows

$$a_k = \frac{1}{2N} \sum_{j=-N+1}^{N} W^{-kj} z_j ,$$

(which is the analogue of (8.5)) with

$$W = e^{2i\pi/2N} \qquad \text{and} \qquad z_j = u(\tilde{x}_j).$$

We have $\quad \tilde{P}_C : C_P^0(I) \to \tilde{S}_N, \quad$ with

$$(u-\tilde{P}_C u, v_N)_N = 0, \qquad \text{for all} \quad v_N \in \tilde{S}_N \quad \text{(analogue of (8.7))}$$

$$(u,v)_N = \frac{1}{2N} \sum_{j=-N+1}^{N} u(\tilde{x}_j) \overline{v(\tilde{x}_j)} .$$

Finally, as the numerical integration formula

$$\frac{1}{2\pi} \int_{-\pi}^{\pi} f(x)dx \approx \frac{1}{2N} \sum_{j=-N+1}^{N} f(\tilde{x}_j),$$

is exact for $f \in S_{2N-1}$, we have the analogue of (8.8) (the proof is left to the reader as an exercise).

We can deduce an error estimate for $\|u-\tilde{P}_C u\|_0$ similar to that established in Theorem 8.1, namely

(8.20) $$\|u-\tilde{P}_C u\|_0 \leq C(1+N^2)^{-r/2} \|u\|_r.$$

Estimation of $\|u-P_C u\|_s$.

Proposition 8.1: <u>For</u> $\sigma < s$, <u>we have the "inverse" inequality</u>

$$\|v_N\|_s \leq (1+N^2)^{(s-\sigma)/2} \|v_N\|_\sigma, \qquad \text{for all } v_N \in S_N.$$

Proof: We have, for $v_N \in S_N$:

$$\|v_N\|_s^2 = \sum_{|m| \leq N} (1+m^2)^s |\hat{v}_m|^2 \leq (1+N^2)^{s-\sigma} \sum_{|m| \leq N} (1+m^2)^\sigma |\hat{v}_m|^2 = (1+N^2)^{s-\sigma} \|v_N\|_\sigma^2,$$

and the result follows.

Corollary 8.2: <u>Suppose</u> $s < r$ <u>is given, then, there exists a constant</u> C <u>such that if</u> $u \in H_P^r(I)$, <u>we have</u>

$$\|u-P_C u\|_s \leq C(1+N^2)^{(s-r)/2} \|u\|_r.$$

Proof: Recalling the identity (8.15) we see that

$$\|u-P_C u\|_s \leq \|u-P_N u\|_s + \|P_C v\|_s.$$

We note that Theorem 6.1 takes care of the first term. We now turn to the second one.

As $P_C v \, \varepsilon \, S_N$, we may apply the inverse inequality established in Proposition 8.1 (with $\sigma = 0$).

$$\| P_C \, v \|_s \, \leqslant \, (1+N^2)^{s/2} \, \| P_C v \|_0 \, ,$$

and the required result follows from (8.16).

9. First-Order Equations – The Collocation Method[1]

We shall now study a semi-discretization more realistic than the Galerkin method which can be applied in the case where the coefficient, a, entering the definition of the operator L

$$Lu \equiv a \frac{\partial u}{\partial x} + \frac{\partial}{\partial x} (au),$$

is not constant.

The approximate problem is then the following.

Find $u_C(t) \in S_N$ such that

$$\frac{\partial}{\partial t} u_C + L_C u_C = 0 \qquad \text{for} > 0$$

(9.1)

$$u_C(0) = P_C u_0.$$

where the operator $L_C : S_N \to S_N$ is defined by

(9.2) $$L_C u \equiv P_C \left(a \frac{\partial u}{\partial x} \right) + \frac{\partial}{\partial x} \left(P_C(a(x)u) \right).$$

We now show that the operator L_C is <u>antisymmetric</u>. Using definition (8.7) of P_C, and (8.8), we have for $u, v \in S_N$

---

[1] also called "pseudo-spectral" method.

(9.3) $(L_C u,v) = \left(a \frac{\partial u}{\partial x}, v\right)_N + \left(\frac{\partial}{\partial x}(P_C(au)),v\right)$ = (integrating by parts)

$$= \left(\frac{\partial u}{\partial x}, av\right)_N - \left(au, \frac{\partial v}{\partial x}\right)_N$$

$$= \left(\frac{\partial u}{\partial x}, P_C(av)\right)_N - \left(u, P_C\left(a \frac{\partial v}{\partial x}\right)\right)_N$$

$$= -\left(u, \frac{\partial}{\partial x} P_C(av)\right) - \left(u, P_C\left(a \frac{\partial v}{\partial x}\right)\right)$$

$$= -(u, L_C v).$$

Since S_N is of finite dimension the equation (9.1) yields a differential system which admits a unique solution

$$u_C \in C^0(0,T ; S_N).$$

Remark 9.1: If a is constant, L_C and L coincide on S_N; as is also true for L and L_N, we deduce that $u_C = u_N$ and with it the equivalence of the Galerkin and collocation methods for this case.

We return to the general case of nonconstant coefficient and establish the following convergence result

Theorem 9.1:   Let $\tau > 1$ and $T > 0$ given, then there exists a constant C such that if $u_0 \in H_P^\tau(I)$ and u is the solution of the equation (7.1), then

$$\|u(t) - u_C(t)\|_0 \leq C(1+N^2)^{\frac{1-\tau}{2}} \|u_0\|_\tau, \qquad \text{for } 0 < t < T.$$

Proof: Let

$$\tilde{u}_N(t) \equiv P_N u(t) \quad \text{and} \quad z(t) = (\tilde{u}_N - u)(t),$$

we have from (7.1)

$$\frac{\partial \tilde{u}_N}{\partial t} + L\tilde{u}_N = \frac{\partial z}{\partial t} + Lz,$$

that is to say

$$\frac{\partial \tilde{u}_N}{\partial t} + L_C \tilde{u}_N = (L_C - L)\tilde{u}_N + \frac{\partial z}{\partial t} + Lz,$$

so by subtracting (9.1), and setting $W_N = \tilde{u}_N - u_C$

$$\frac{\partial W_N}{\partial t} + L_C W_N = (L_C - L)\tilde{u}_N + \frac{\partial z}{\partial t} + Lz.$$

Multiplying this relation by W_N in terms of the scalar product $(\cdot, \cdot)$, we have

$$\left(\frac{\partial W_N}{\partial t}, W_N\right) + (L_C W_N, W_N) = ((L_C - L)\tilde{u}_N, W_N) + \left(\frac{\partial z}{\partial t}, W_N\right) + (Lz, W_N).$$

From the antisymmetry of L_C, we deduce that, (passing to the real parts)

$$\|W_N\|_0 \frac{d}{dt} \|W_N\|_0 \equiv \frac{1}{2} \frac{d}{dt} \|W_N\|_0^2 = \text{Re}((L_C - L)\tilde{u}_N, W_N) + \text{Re}\left(\frac{\partial z}{\partial t}, W_N\right) + \text{Re}(Lz, W_N).$$

The Schwartz inequality then yields

$$\frac{d}{dt} \|W_N\|_0 < \|(L_C - L)\tilde{u}_N\|_0 + \left\|\frac{\partial z}{\partial t}\right\|_0 + \|Lz\|_0$$

so integrating between 0 and T we find

$$(9.4) \qquad \|W_N(t)\|_0 < \|W_N(0)\|_0 + t \underset{t\epsilon[0,T]}{\text{Sup}} \left(\|(L_C-L)\tilde{u}_N\|_0 + \|\frac{\partial z}{\partial t}\|_0 + \|Lz\|_0 \right).$$

We will now obtain an estimate for the Sup term in (9.4). First consider

$$(L_C-L)\tilde{u}_N = (P_C-I)a \frac{\partial \tilde{u}_N}{\partial x} + \frac{\partial}{\partial x} (P_C-I)a\tilde{u}_N.$$

Let

$$y(t) = a \frac{\partial \tilde{u}_N}{\partial x} (t);$$

as $a \epsilon C_P^\infty(I)$, we have

$$\|y(t)\|_{\tau-1} < C\|\frac{\partial \tilde{u}_N(t)}{\partial x}\|_{\tau-1} < C\|\tilde{u}_N(t)\|_\tau.$$

Furthermore, from the definition of the norm $\|\cdot\|_r$ given in section 6, it follows immediately that

$$(9.5) \qquad \|\tilde{u}_N(t)\|_\tau < \|u(t)\|_\tau$$

since $\tilde{u}_N(t) = P_N u(t)$.

We deduce then from Theorem (8.1) (applied with $r = \tau-1$) that

$$\|y-P_C y\|_0 < C(1+N^2)^{-(\tau-1)/2}\|y\|_{\tau-1}$$

$$< C(1+N^2)^{-(\tau-1)/2}\|u\|_\tau.$$

We have

$$\| (L_C - L)\widetilde{u}_N \|_0 \ \leq\ \| (P_C - I)y \|_0 + \| (P_C - I)a\widetilde{u}_N \|_1$$

$$\leq\ C(1+N^2)^{(1-\tau)/2} \| u \|_\tau + \| (P_C - I)a\widetilde{u}_N \|_1.$$

Applying Corollary 8.2 with $s = 1$ and $r = \tau$ we find that

$$\| (P_C - I)a\widetilde{u}_N \|_1 \ \leq\ C(1+N^2)^{\frac{1-\tau}{2}} \| a\widetilde{u}_N \|_\tau,$$

Using inequality (9.5) we have

$$\| a\widetilde{u}_N \|_\tau \ \leq\ C\| \widetilde{u}_N \|_\tau \ \leq\ C\| u \|_\tau.$$

This establishes that

$$(9.6) \qquad \| (L_C - L)\widetilde{u}_N(t) \|_0 \ \leq\ C(1+N^2)^{\frac{1-\tau}{2}} \| u(t) \|_\tau.$$

As

$$z(t) = -(I - P_N)u(t),$$

it follows that

$$(9.7) \qquad \| Lz \|_0 \ \leq\ C\| z \|_1 = C\| u - P_N u \|_1 \ \leq\ C(1+N^2)^{\frac{1-\tau}{2}} \| u(t) \|_\tau$$

$$(9.8) \qquad \| \frac{\partial z}{\partial t} \|_0 = \| (I - P_N)\frac{\partial u}{\partial t} \|_0 \ \leq\ C(1+N^2)^{\frac{1-\tau}{2}} \| \frac{\partial u}{\partial t} \|_{\tau-1} \ \leq\ C(1+N^2)^{\frac{1-\tau}{2}} \| u \|_\tau,$$

where the last inequality is gotten by noting

$$\frac{\partial u}{\partial t} = -Lu, \qquad \text{and} \qquad \|Lu\|_{\tau-1} \leq C\|u\|_{\tau}.$$

Therefore

$$\sup_{t \in [0,T]} \left(\|(L_C - L)\tilde{u}_N\|_0 + \|\frac{\partial z}{\partial t}\|_0 + \|Lz\|_0 \right) \leq C(1-N^2)^{\frac{1-\tau}{2}} \|u_0\|_{\tau}.$$

We also have

$$\|W_N(0)\|_0 = \|P_C u - P_N u\|_0 = \|P_C(u-P_N u)\|_0 \leq C(1+N^2)^{-\frac{\tau}{2}} \|u_0\|_{\tau},$$

(see (8.16)).

From (9.4) then we have

$$\|W_N(t)\|_0 \leq (1+Ct)(1+N^2)^{\frac{1-\tau}{2}} \|u_0\|_{\tau}, \qquad 0 \leq t \leq T,$$

and the result follows since

(9.9) $$\|u(t) - u_C(t)\|_0 \leq \|u(t)-\tilde{u}_N(t)\|_0 + \|W_N(t)\|_0.$$

Q.E.D.

Remark 9.2: The preceding result shows that if $u \in H_p^{\tau}(I)$, with $\tau > 1$, we have an error estimate of order $O(N^{1-\tau})$ which is identical to that obtained for the Galerkin method (see Remark 7.1).

10. Time Discretization Schemes:

Suppose A is a $M \times M$ matrix and $U(t) \in \mathbb{C}^M$ is the solution of the differential system

$$\frac{d}{dt} U + AU = 0,$$

(10.1)

$$U(0) = U_0.$$

We can discretize (10.1) by either implicit or explicit schemes. The former correspond to the approximation of the true exponential solution by some rational fraction, the latter by polynomials.

For example, the scheme

$$U^{n+1} = (I + \Delta tA)^{-1} U^n,$$

is an implicit scheme, since for each iteration there is a linear system to solve (a matrix to invert). As

$$U\big((n+1)\Delta t\big) = e^{-\Delta tA} U(n\Delta t),$$

and as $(I + \Delta tA)^{-1}$ is an approximation of $e^{-\Delta tA}$ for Δt sufficiently small, this algorithm converges.

In contrast the schemes

(10.2)
$$U^{n+1} = P_J(\Delta tA)U^n$$

where

$$(10.3) \qquad P_J(\tau) = \sum_{j=0}^{J} \frac{(-\tau)^j}{j!} ,$$

are explicit, because there is no linear system to solve at each iteration.

They are convergent since the polynomials $P_J(\tau)$ constitute approximations (of order J) to the exponential $e^{-\tau}$ when τ is sufficiently small (and thus the matrices $P_J(\Delta tA)$ approximate $e^{-\Delta tA}$).

We do not assert a priori that the explicit schemes have a big advantage in terms of efficiency over the implicit schemes for the general case of a full matrix A. But in the case of the collocation method studied in section 9 we saw that the product of a vector U^n by the matrix A can be evaluated rapidly, using the Fast Fourier Transform (see Remark 8.1). Let us examine these schemes now in some detail.

The first question that presents itself is that of stability. For the scheme to converge, it is not sufficient that the matrix $P_J(\Delta t)$ be an approximation to the exponential; it is further required that the spectral radius be less than 1, otherwise the sequence U^n generated by the algorithm (10.2) will increase exponentially.

Whether this is so depends on the <u>spectrum</u> of the matrix A.

Proposition 10.1: <u>The differential system (9.1) is of the type (10.1)</u> <u>with an antisymmetric matrix</u> A <u>of order</u> $M \times M$ <u>with</u> $M = 2N+1$.

<u>Proof</u>: Suppose

$$\psi_k = \frac{1}{2N+1} \sum_{|n| \leqslant N} W^{-nk} e^{inx},$$

where W is (as in section 8) the principal root of order (2N+1) of unity;
we have shown in section 8) that

(10.4) $\psi_k(x_j) = \delta_{jk}$ for $|j|$, $|k| \leq N$ and $\psi_k \in S_N$.

Therefore, for all $u \in S_N$ we have

$$u(x) = \sum_{|k| \leq N} \psi_k(x)u(x_k).$$

We set $U_k = u(x_k)$, $|k| \leq N$ so that $U = (U_k) \in \mathbb{C}^{2N+1}$.

Further, the functions ψ_k are orthogonal; in fact, according to (8.8),
we have

(10.5) $(\psi_k, \psi_\ell) = (\psi_k, \psi_\ell)_N = \frac{1}{2N+1} \sum_{|j| \leq N} \psi_k(x_j) \overline{\psi_\ell(x_j)} = \frac{1}{2N+1} \delta_{\ell k}.$

Letting

$$u_C(t) = \sum_{|k| \leq N} U_k(t)\psi_k(x),$$

we may write (9.1) in the equivalent form

$$\left(\frac{\partial}{\partial t} u_C, \psi_\ell\right) + \left(L_C u_C, \psi_\ell\right) = 0, \qquad |\ell| \leq N,$$

that is to say

$$\sum_{|k| \leq N} \left[(\psi_k, \psi_\ell) \frac{dU_k}{dt} + (L_C\psi_k, \psi_\ell)U_k \right] = 0.$$

Then using account (10.5), we get

$$\frac{dU_\ell}{dt} + (2N+1) \sum_{|k| \leqslant N} (L_C \psi_k, \psi_\ell) U_k = 0,$$

which shows that (9.1) is indeed a differential system of the form (10.1) with a matrix A of order M×M (where M = 2N+1) defined by

$$A = (a_{\ell k})_{|k|, |\ell| \leqslant N},$$

and

(10.6)
$$a_{\ell k} = (2N+1)(L_C \psi_k, \psi_\ell).$$

The antisymmetry (9.3) of L_C shows that the matrix A is anti-symmetric.   Q.E.D.

Remark 10.1: The Galerkin method amounts to solving

$$\left(\frac{\partial u_N}{\partial t}, v_N\right) + \left(a \frac{\partial u_N}{\partial x}, v_N\right) - \left(a u_N, \frac{\partial v_N}{\partial x}\right) = 0, \qquad \text{for all } v_N \in S_N,$$

without numerical integration.

The collocation method amounts to solving

(10.7)
$$\left(\frac{\partial u_C}{\partial t}, v_N\right) + \left(a \frac{\partial u_C}{\partial x}, v_N\right)_N - \left(a u_C, \frac{\partial v_N}{\partial x}\right)_N = 0,$$

where a form of numerical summation (8.9) has been used to evaluate the integrals which are generally impossible to evaluate exactly.

Remark 10.2: In practice, to evaluate the product of the matrix A by a given vector U, we first note that the coefficients $a_{\ell k}$ of the matrix A given by (10.6) may equally well be written using (8.8) as

$$a_{\ell k} = (2N+1)(L_C \psi_k, \psi_\ell)_N ,$$

that is to say

$$a_{\ell k} = \sum_{|j| \leq N} ((L_C \psi_k)(x_j)) \overline{\psi_\ell(x_j)} = (L_C \psi_k)(x_\ell)$$

(In this form in which the skew symmetry of A is less obvious). Suppose then that

$$(10.8) \qquad u(x) = \sum_{|k| \leq N} \psi_k(x) U_k .$$

In order to calculate

$$(AU) \equiv \sum_k a_{\ell k} U_k = \sum_k (L_C \psi_k)(x_\ell) U_k = (L_C u)(x_\ell),$$

it is sufficient to calculate $L_C u$ at the points x_ℓ given the values of the function u at the same points.

Using (9.2) it is possible therefore to proceed in the following fashion

1. Given $u(x_j) = U_j$, we calculate the Fourier coefficients a_n of u (see Corollary 8.1) with the help of Fast Fourier Transform (FFT).

2. We deduce the Fourier coefficients of $v = \dfrac{\partial u}{\partial x}$ by means of the formula $\hat{v}_n = i n a_n$.

3. Using the inverse FFT we obtain the values $v(x_\ell)$ of $\frac{\partial u}{\partial x}$ at the points x_ℓ and then the values of $a\frac{\partial u}{\partial x}$ at x_ℓ.

In order to evaluate $(L_C u)(x_\ell)$ it remains to calculate $(\frac{\partial}{\partial x} P_C(au))(x_\ell)$. For that (in an analogous manner)

(i) We calculate by multiplication

$$W(x_j) = a(x_j)u(x_j) \equiv \bigl(P_C(au)\bigr)(x_j)$$

(ii) next (via an FFT) the Fourier coefficients

$$\hat{W}_n \quad \text{from} \quad W \equiv P_C(au).$$

(iii) we have therefore by multiplication (cf. 2 above) the Fourier coefficients of $\frac{\partial W}{\partial x}$; then by an inverse FFT the values of $\frac{\partial W}{\partial x}(x_\ell)$ and finally $(L_C u)(x_\ell)$.

As the FFT has an operation count of $O(M \log_2 M)$ (see section 8), we infer that the calculation of the product AU costs only $O(M\log_2 M)$ operations. (Instead of the $O(M^2)$ operations needed if we calculated the product directly).

We shall then use discretization schemes for differential systems of the type (10.1) which are efficient when the matrix A is antisymmetric.

For example we may use the three-level leap-frog scheme (of second-order) (see e.g., Richtmyer-Morton [14])

$$(10.9) \qquad \frac{U^{n+1} - U^{n-1}}{2\Delta t} + AU^n = 0,$$

or one of the schemes (10.2) provided that the following condition is satisfied

(10.10)   <u>there exists</u> $\delta > 0$ <u>such that</u> $|\tau| < \delta, \tau \in \mathbf{R}$ <u>implies</u> $|P_J(i\tau)| < 1$.

In fact, as A is antisymmetric it is diagonalizable by a unitary matrix Q, where $D = Q^*AQ$ diagonal and pure imaginary.

Let
$$\lambda_j = d_{jj} \qquad \text{and} \qquad V^n = Q^*U^n;$$

we may write
$$QV^{n+1} = P_J(\Delta tA)QV^n$$

$$V^{n+1} = Q^*P_J(\Delta tA)QV^n$$

$$V^{n+1} = P_J(\Delta tD)V^n$$

whence
$$V^{n+1} = P_J(\Delta t\lambda_j)V^n_j \qquad\qquad |j| < N.$$

The condition (10.10) shows that if Δt is chosen sufficiently small so that $|\Delta t\lambda_j| < \delta$, for all j, then

$$V^n_j = \left(P_J(\Delta t\lambda_j)\right)^n V^0_j$$

is bounded

$$|v_j^n| < |v_j^0|$$

and this proves the stability of the scheme.

It is well known that the condition (10.10) is satisfied, e.g., if $J = 3,4,7$ or 8.

For example, for $J = 3$ we have

$$P_J(i\tau) = 1 - i\tau - \frac{\tau^2}{2} - \frac{(i\tau)^3}{6} = 1 - \frac{\tau^2}{2} - i\left(\tau - \frac{\tau^3}{6}\right)$$

$$|P_J(i\tau)|^2 = \left(1 - \frac{\tau^2}{2}\right)^2 + \left(\tau - \frac{\tau^3}{6}\right)^2 = 1 - \frac{\tau^4}{12} + \frac{\tau^6}{36} .$$

Q.E.D.

To finish the analysis of stability, it remains only to find an upper bound for the eigenvalues of the matrix A.

Proposition 10.2: If $\lambda \in \mathrm{Sp}(A)$ for A defined by (10.6), then $|\lambda| < CN$.

Proof: Suppose $\lambda \in \mathrm{Sp}(A)$ and U is an eigenvector so that $AU = \lambda U$. From (10.5) and (10.6) we have by setting $u = \sum_{|k| < N} U_k \psi_k$ (as in (10.8)

$$\lambda(2N+1)(u,u) = \lambda U^* U = U^* A U = (2N+1)(L_c u, u),$$

i.e., assuming $\|u\|_0 = 1$,

$$\lambda = (L_c u, u).$$

Now (see (10.7)) (denoting by Im the imaginary part of a complex number)

$$(L_c u, u) = \left(a \frac{\partial u}{\partial x}, u\right)_N - \left(au, \frac{\partial u}{\partial x}\right)_N = 2 \ \mathrm{Im}\left(a \frac{\partial u}{\partial x}, u\right)_N \ ,$$

and

$$\left|\left(a \frac{\partial u}{\partial x}, u\right)_N\right| \leqslant \max_j |a(x_j)| \ \|\frac{\partial u}{\partial x}\|_{0,N} \ \|u\|_{0,N} \ ,$$

where $\|u\|_{0,N}$ is the norm of u associated with the scalar product defined in (8.6). As $u \in S_N$, we have

$$\|\frac{\partial u}{\partial x}\|_{0,N} = \|\frac{\partial u}{\partial x}\|_0 = \left(\sum_{|n| \leqslant N} n^2 |\hat{u}_n|^2\right)^{1/2} \leqslant N\left(\sum_{|n| \leqslant N} |\hat{u}_n|^2\right)^{1/2} = N\|u\|_0 = N$$

consequently, $|\lambda| \leqslant 2N \max_j |a(x_j)|$ and this furnishes an evaluation of the constant C.

Corollary 10.1: <u>If</u> $\Delta t < \frac{\delta}{CN}$ <u>where</u> δ <u>is the constant introduced in (10.10) and</u> C <u>is the constant of the Proposition 10.2, the iterative method (10.2) is stable; we have</u>

$$\|U^{n+1}\|_0 \leqslant \|U^n\|_0,$$

<u>where</u> $\|U\|_0 \stackrel{\text{def}}{=} \left(\frac{1}{2N+1} \sum_{|k| \leqslant N} |U_k|^2\right)^{1/2} \ .$

<u>Error Estimate</u>

Let $g \in S_N$ be given, we define

$$U^0(g) = g$$

$$U^{n+1}(g) = P_J(\Delta t L_C)U^n(g), \qquad n \geqslant 0$$

and we also define $u_C(t;g)$ to be the solution of the equation

$$\frac{\partial u_C}{\partial t} + L_C u_C = 0$$

$$u_C(0;g) = g.$$

According to (10.10), we have, for Δt chosen as in the Corollary 10.1,

(10.11)
$$\|U^n(g)\|_0 \leqslant \|g\|_0.$$

On the other hand the skew symmetry of L_C shows that

$$Re\left(\frac{\partial u_C}{\partial t}, u_C\right) = 0 \quad > \quad \frac{1}{2}\frac{d}{dt}\|u_C(t)\|^2 = 0,$$

and therefore that

(10.12)
$$\|u_C(t;g)\|_0 = \|g\|_0.$$

Setting

(10.13)
$$E_j(g) \equiv U^j(g) - u_C(t_j,g),$$

where $t_j \equiv j\Delta t$; we note that E_j is linear in g.

(This results from the commutative properties of the resolvent of an operator and the semi-group associated with this operator, (see Kato [10]).

Let $\lambda_k \in \mathbb{C}$ and $\Phi_k \in S_N$ denote the eigenvectors and eigenfunctions of L_C. Set

$$u_C(t_j, g) = \sum_k a_k \Phi_k$$

$$u_C(t_j, f) = \sum_k \beta_k \Phi_k$$

$$\left(P_j(\Delta t L_C) - \exp(-\Delta t L_C)\right) u_C(t_j, f) = \sum_k \gamma_k \Phi_k.$$

We have

$$\beta_k = \frac{\alpha_k}{(1+\lambda_k)^m}$$

and

$$\gamma_k = \left[P_J(\Delta t \lambda_k) - \exp(-\Delta t \lambda_k)\right]\beta_k = \frac{\alpha_k}{(1+\lambda_k)^m} \sum_{\ell > J} \frac{(-\Delta t \lambda_k)^\ell}{\ell !}$$

$$= \alpha_k \frac{\Delta t^m \lambda_k^m}{(1+\lambda_k)^m} \sum_{\ell' > J-m} \frac{(-\Delta t \lambda_k)^{\ell'}}{(\ell'+m)!} .$$

As by hypothesis $|\Delta t \lambda_k| < \delta$, we have

$$\left| \sum_{\ell' > J-m} \frac{(-\Delta t \lambda_k)^{\ell'}}{(\ell'+m)!} \right| < \sum_{\ell' > 0} \frac{\delta^{\ell'}}{(\ell')!} = e^\delta,$$

therefore

$$|\gamma_k| < |\alpha_k| \Delta t^m e^\delta.$$

Thus ST satisfies

$$ST \equiv \left(\sum_k |\gamma_k|^2 \right)^{1/2} \leqslant \Delta t^m e^\delta \left(\sum_k |\alpha_k|^2 \right)^{1/2} = \Delta t^m e^\delta \| u_C(t_j,g) \|_0 \ .$$

We conclude (using (10.12))

$$\| E_{j+1}(f) \|_0 \leqslant \| E_j(f) \|_0 + \Delta t^m e^\delta \| g \|_0 \ ,$$

whence (ii) with $C = t_0 e^\delta$ by summation from $j = 0$ to n, and using the hypothesis that $n\Delta t \leqslant t_0$.

Q.E.D.

We are now in a position to establish the principal result.

Theorem 10.1: <u>If $u_0 \varepsilon H_p^{\tau+J}(I)$, we have the error estimate</u>

$$\| u(t_n) - U^n \|_0 \leqslant C(N^{1-\tau} + \Delta t^J),$$

<u>where $U^n = U^n(P_C u_0)$ is the solution at time $t_n = n\Delta t$ for the completely discretized problem.</u>

<u>Proof</u>: We establish (by induction on J) the following identity

$$u \equiv \sum_{j=1}^{J+1} T_C^j (L_C - L_N)(I+L_N)^{j-1} u + T_C^{J+1} (I+L_N)^{J+1} u,$$

for all $u \varepsilon S_N$.

We infer from the linearity of the operator E_n defined in (10.13) that

$$E_n(P_C u_0) = E_n(P_C u_0 - P_N u_0) + E_n(P_N u_0)$$

$$(10.14) \qquad E_n(P_N u_0) = \sum_{j=1}^{J+1} E_n\left(T_C^j(L_C-L_N)(I+L_N)^{j-1} P_N u_0\right)$$

$$+ E_n\left(T_C^{J+1}(I+L_N)^{J+1} P_N u_0\right).$$

Applying result (ii) of Lemma 10.1, we have for $j=1,\cdots,J+1$

$$\| E_n\left(T_C^j(L_C-L_N)(I+L_N)^{j-1} P_N u_0\right) \|_0 \leq C\Delta t^{j-1} \| (L_C-L_N)(I+L_N)^{j-1} P_N u_0 \|_0.$$

From the definition of L_N, we have, for $v \in S_N$

$$\| (L_C-L_N)v \|_0 \leq \| (L_C-L)v \|_0 + \| (I-P_N)Lv \|_0$$

$$\leq C(1+N^2)^{\frac{1-\tau}{2}} \| v \|_\tau + (1+N^2)^{\frac{1-\tau}{2}} \| Lv \|_{\tau-1},$$

where we have applied a variant of (9.6) for the first term and Theorem 6.1 to the second term.

As $\| Lv \|_{\tau-1} \leq C\| v \|_\tau$ (since the coefficient a is smooth) we have

$$\| (L_C-L_N)v \|_0 \leq C(1+N^2)^{\frac{1-\tau}{2}} \| v \|_\tau.$$

Finally, supposing $v = (I+L_N)^{j-1} P_N u_0$, we have from the continuity of L from $H_P^s(I)$ in $H_P^{s+1}(I)$

$$\| v \|_\tau \leq C\| P_N u_0 \|_{\tau+j-1} \leq C\| u_0 \|_{\tau+j-1}.$$

Then the last term needed to estimate in (10.14) is

$$\| E_n(T_C^{J+1}(I+L_N)^{J+1}P_N u_0)\| \leq C\Delta t^J \|u_0\|_{J+1}.$$

We have then

$$\| E_n(P_N u_0)\| \leq \sum_{j=1}^{J+1} C\Delta t^{j-1}(1+N^2)^{\frac{1-\tau}{2}} \|u_0\|_{\tau+j-1} + C\Delta t^J \|u_0\|_{J+1}$$

and

$$\| E_n(P_C u_0)\|_0 \leq \| E_n(P_C u_0 - P_N u_0)\|_0 + \| E_n(P_N u_0)\|_0$$

$$\leq \| P_C u_0 - u_0\|_0 + \| u_0 - P_N u_0\|_0 + \| E_n(P_N u_0)\|_0$$

$$\leq C(1+N^2)^{\frac{1-\tau}{2}} \|u_0\|_{\tau-1} + \| E_n(P_N u_0)\|_0,$$

that is to say

$$\| E_n(P_C u_0)\| \leq C\left((1+N^2)^{\frac{1-\tau}{2}} + \Delta t^J\right)\|u_0\|_{\tau+J},$$

we conclude by noting that if $g = P_C u_0$, $\| E_n(P_C u_0)\|_0$ gives the error between the solution of the semi-discrete problem and that of the fully discrete problem. As the error between $u(t_n)$ and $u_C(t_n)$ is of order $N^{1-\tau}$ according to section 9, we have the desired result.

Remark 10.3:

1. The error estimate established in Theorem 10.1 requires strong

regularity for the initial solution u_0, (and hence the exact solution $u(t)$).

For the case of the weaker regularity $u_0 \in H_p^\tau(I)$, we can prove, in the same manner, convergence of order $O(\Delta t^J)$ of U_n to $u_c(t_n)$ as $\Delta t \to 0$ but the constant introduced in this case depends a priori on N.

2. In practice, as the time step is limited by the stability condition established in Corollary 10.1, it is not useful to take the order J of the schemes to be very high ($J = 3$ seems a reasonable choice). We might as well use the leap-frog scheme which is second order accurate and requires only the product of the matrix A by a vector at each iteration.

11. An Advection-Diffusion Equation

We consider now the parabolic equation

i) $\quad \dfrac{\partial u}{\partial t} + Tu = 0, \qquad\qquad\qquad t \geqslant 0, \quad x \in I,$

ii) $\quad u(0,x) = u_0(x) \qquad\qquad\qquad$ (initial condition),

iii) $\quad u(t,-\pi) = u(t,\pi), \dfrac{\partial u}{\partial x}(t,-\pi) = \dfrac{\partial u}{\partial x}(t,\pi)$ (periodicity condition),

where the operator T is given by

$$T = \varepsilon A + L,$$

where A is the diffusion operator

(11.2)
$$A = -\frac{\partial}{\partial x} b(x) \frac{\partial}{\partial x} + e(x),$$

and L is the advection operator

$$Lu = a \frac{\partial u}{\partial x} + \frac{\partial}{\partial x} (au).$$

The coefficients a, b, and e of the operator T assumed to be real, periodic, and regular. $\varepsilon > 0$ is a real number.

We shall examine the dependence of the solution, u_ε, on ε.

We suppose that there exist constants $\beta > 0$ and $\gamma \in \mathbb{R}$ such that

(11.3)
$$b(x) \geqslant \beta, \quad e(x) \geqslant -\gamma, \quad \text{for all } x \in I.$$

This means that

$$(Au,u) \geqslant \beta \|\frac{\partial u}{\partial x}\|_0^2 - \gamma \|u\|_0^2, \quad \text{for all } u \in H_P^2(I)$$

(11.4)

$$(Au,u) \geqslant \beta \|u\|_1^2 - (\gamma+\beta) \|u\|_0^2.$$

The existence of a solution u_ε of (11.1) for $\varepsilon > 0$ follows then from the classical results on parabolic problems, (see e.g., Lions-Magenes [11]).

We will confine our attention to establishing an a priori estimate for u_ε.

Theorem 11.1: Let $\Delta > 0$ and $s \geqslant 0$ given; then there exists a positive constant C such that for all $\varepsilon > 0$ and $t \in [0,\Delta]$ we have the inequality

$$\| u_\varepsilon(t) \|_s \leq C \| u_0 \|_s .$$

Proof: In a manner analogous to the proof of Theorem 7.1, we introduce
the operator

$$\Lambda^s : H_P^s(I) \to L^2(I),$$

such that

$$\| \Lambda^s u \|_0 = \| u \|_s .$$

Recall that Λ^s is an operator (pseudo-differential, in the general case
where s is not even integer) of order s. We have then

$$\frac{d}{dt} \| u_\varepsilon(t) \|_s^2 = \left(\Lambda^s(-Tu_\varepsilon), \Lambda^s u_\varepsilon \right) + \left(\Lambda^s u_\varepsilon, \Lambda^s(-Tu_\varepsilon) \right)$$

$$= -\left((L+L^*)\Lambda^s u_\varepsilon, \Lambda^s u_\varepsilon \right) - 2\mathrm{Re}\left(K u_\varepsilon, \Lambda^s u_\varepsilon \right)$$

$$- 2\varepsilon \left(A\Lambda^s u_\varepsilon, \Lambda^s u_\varepsilon \right) - \varepsilon \left([\Lambda^s, A] u_\varepsilon, \Lambda^s u_\varepsilon \right) - \varepsilon \left(\Lambda^s u_\varepsilon, [\Lambda^s, A] u_\varepsilon \right),$$

where $K \equiv [\Lambda^s, L] \equiv \Lambda^s L - L\Lambda^s$ denotes the commutator of Λ^s and L.

In order to get an upper bound on u_ε, we can use successively the
antisymmetry of L, the fact that K is of order s, the coercivity of A
(see (11.4)), and finally the fact that the operator $[\Lambda^s, A]$ is of order
s+1, to yield the result that

$$\| [\Lambda^s, A] u_\varepsilon \|_0 \le C_1 \| u_\varepsilon \|_{s+1} .$$

We obtain

$$(11.5) \qquad \frac{d}{dt} \| u_\varepsilon(t) \|_s^2 \le 2 \big(C + \varepsilon (\gamma + \beta) \big) \| u_\varepsilon \|_s^2 - 2 \varepsilon \beta \| \Lambda^s u_\varepsilon \|_1^2 + 2 C_1 \varepsilon \| u_\varepsilon \|_{s+1} \| u_\varepsilon \|_s .$$

Then using the inequality

$$2 \| u_\varepsilon \|_{s+1} \| u_\varepsilon \|_s \le \alpha \| u_\varepsilon \|_{s+1}^2 + \frac{1}{\alpha} \| u_\varepsilon \|_s^2 ,$$

with α taken equal to $\dfrac{\beta}{C_1}$, (noting that $\| \Lambda^s u_\varepsilon \|_1 = \| u_\varepsilon \|_{s+1}$)) we find that

$$\frac{d}{dt} \| u_\varepsilon(t) \|_s^2 \le C_2 \| u_\varepsilon \|_s^2$$

with

$$C_2 \equiv 2 \big(C + \varepsilon (\gamma + \beta) \big) + \frac{C_1^2 \varepsilon}{\beta} .$$

Thus

$$\| u_\varepsilon(t) \|_s^2 \le e^{C_2 t} \| u_0 \|_s^2 ,$$

and the result follows by, noting that C_2 is bounded independently of ε.

Q.E.D.

The Semi-discrete Problem

We introduce the operator A_C defined by

(11.6)
$$A_C u = - \frac{\partial}{\partial x} \left(P_C (b \frac{\partial u}{\partial x}) \right) + P_C (eu),$$

which is an operator from S_N to itself.

Set

(11.7)
$$T_C = \varepsilon A_C + L_C$$

where L_C is the operator, studied in sections 9 and 10, defined by

$$L_C u = P_C \left(a \frac{\partial u}{\partial x} \right) + \frac{\partial}{\partial x} P_C (au).$$

The semi-discrete problem is then to find $u_C(t) \in S_N$ satisfying

(11.8)

(i) $\qquad \frac{\partial}{\partial t} u_C + T_C u_C = 0$

(ii) $\qquad u_C(0) = P_C(u_0).$

Lemma 11.1: The operator T_C defined in (11.7) satisfies the coercivity inequality

$$Re(T_C u, u) \geqslant \varepsilon \beta \left\| \frac{\partial u}{\partial x} \right\|_0^2 - \varepsilon \gamma \| u \|_0^2, \qquad \text{for all } u \in S_N.$$

Proof: As $Re(L_C u, u) = 0$ from (9.3), it suffices to establish that

$$\text{Re}(A_C u, u) \geqslant \beta \| \frac{\partial u}{\partial x} \|_0^2 - \gamma \| u \|_0^2.$$

Now, we have for $u \in S^N$

$$
\begin{aligned}
(A_C u, u) &= \left(-\frac{\partial}{\partial x}\left(P_C\left(b\,\frac{\partial u}{\partial x}\right)\right), u\right) + (P_C(eu), u) \\[2mm]
&= \left(P_C\left(b\,\frac{\partial u}{\partial x}\right), \frac{\partial u}{\partial x}\right) + (eu, u)_N \\[2mm]
&= \left(b\,\frac{\partial u}{\partial x}, \frac{\partial u}{\partial x}\right)_N + (eu, u)_N \\[2mm]
&= \frac{1}{2N+1} \sum_{|j|<N} \left(b(x_j)|u'(x_j)|^2 + e(x_j)|u(x_j)|^2\right) \\[2mm]
&\geqslant \frac{1}{2N+1} \sum_{|j|<N} \left(\beta |u'(x_j)|^2 - \gamma |u(x_j)|^2\right) \\[2mm]
&= \beta \| \frac{\partial u}{\partial x} \|_0^2 - \gamma \| u \|_0^2,
\end{aligned}
$$

since $u \in S_N$ (and the fact that the numerical integration formula (8.9) is exact for $f \in S_{2N}$).

Q.E.D.

We study now the error between the solution u of the continuous problem (11.1) and the solution u_C of the discrete problem (11.8). (For simplicity, we shall drop subscript ε.)

Theorem 11.2: <u>Let</u> $\tau > 1$ <u>and</u> $\Delta > 0$ <u>be given; then there exists a</u> <u>constant</u> C <u>(independent of</u> ϵ) <u>such that if</u> $u_0 \in H_P^{\tau+1}(I)$, <u>we have the</u> <u>error estimate:</u>

$$\| u(t) - u_C(t) \|_0 \leq C(1+N^2)^{\frac{1-\tau}{2}} \left(\| u_0 \|_{\tau-1} + \left(\| u_0 \|_\tau^2 + (\| u_0 \|_\tau + \epsilon \| u_0 \|_{\tau+1})^2 \right)^{1/2} \right),$$

<u>for all</u> $t \in [0,\Delta]$.

<u>Proof</u>: To simplify the calculations we suppose $e \equiv 0$ (and hence $\gamma = 0$) (the general case is left to the reader as an exercise).

Suppose

$$\tilde{u}_N(t) = P_N u(t) \qquad \text{and} \qquad z(t) = \tilde{u}_N(t) - u(t).$$

We have from (11.1)

$$\frac{\partial \tilde{u}_N}{\partial t} + T_C \tilde{u}_N = (T_C - T)\tilde{u}_N + \frac{\partial z}{\partial t} + Tz.$$

Letting $W_N = \tilde{u}_N - u_C$, and subtracting (11.8), we deduce that

(11.9)
$$\frac{\partial W_N}{\partial t} + T_C W_N = (T_C - T)\tilde{u}_N + \frac{\partial z}{\partial t} + Tz.$$

Taking an inner product with W_N, and taking the real part, we find that (applying Lemma 11.1)

$$\frac{1}{2} \frac{d}{dt} \| W_N \|^2 + \epsilon \beta \| \frac{\partial W_N}{\partial x} \|_0^2 \leq Z_1 + Z_2 + Z_3,$$

where

$$Z_1 \equiv \mathrm{Re}\big((T_C - T)\tilde{u}_N, W_N\big)$$

$$Z_2 \equiv \mathrm{Re}\big(\frac{\partial z}{\partial t}, W_N\big)$$

$$Z_3 \equiv \mathrm{Re}(TZ, W_N).$$

Let us first find an upper bound for Z_1; we have

$$Z_1 = \varepsilon\, \mathrm{Re}\big((A_C - A)\tilde{u}_N, W_N\big) + \mathrm{Re}\big((L_C - L)\tilde{u}_N, W_N\big)$$

(11.10) $\quad \big((A_C - A)\tilde{u}_N, W_N\big) = \big(-\frac{\partial}{\partial x} P_C b \frac{\partial \tilde{u}_N}{\partial x}, W_N\big) - \big(-\frac{\partial}{\partial x} b \frac{\partial \tilde{u}_N}{\partial x}, W_N\big)$

$$= \big((P_C - I)b\frac{\partial \tilde{u}_N}{\partial x}, \frac{\partial W_N}{\partial x}\big) \le \frac{1}{2\beta}\|(P_C - I)b\frac{\partial \tilde{u}_N}{\partial x}\|^2 + \frac{\beta}{2}\|\frac{\partial W_N}{\partial x}\|_0^2 ,$$

and

$$\mathrm{Re}\big((L_C - L)\tilde{u}_N, W_N\big) \le \frac{\theta}{2}\|W_N\|_0^2 + \frac{1}{2\theta}\|(L_C - L)\tilde{u}_N\|_0^2 ,$$

whence

$$Z_1 \le \frac{\varepsilon}{2\beta}\|(P_C - I)b\frac{\partial \tilde{u}_N}{\partial x}\|_0^2 + \frac{1}{2\theta}\|(L_C - L)\tilde{u}_N\|_0^2 + \frac{\varepsilon\beta}{2}\|\frac{\partial W_N}{\partial x}\|_0^2 + \frac{\theta}{2}\|W_N\|_0^2.$$

Now, if $y(t) = b\frac{\partial \tilde{u}_N}{\partial x}(t)$, we have in a manner analogous to the proof of Theorem 9.1:

(11.11) $$\|y - P_C y\|_0^2 \le C(1 + N^2)^{1-\tau}\|u(t)\|_\tau^2,$$

and according to (9.6)

$$\| (L_C - L)\tilde{u}_N(t)\|_0^2 \leqslant C(1+N^2)^{1-\tau} \|u(t)\|_\tau^2,$$

whence

$$z_1 \leqslant C\Big(\frac{1}{\beta} + \frac{1}{\theta}\Big)(1+N^2)^{1-\tau}\|u(t)\|_\tau^2 + \frac{\beta}{2} \|\frac{\partial W_N}{\partial x}\|_0^2 + \frac{\theta}{2} \|W_N\|_0^2 \ .$$

Moving onto z_2, we have

$$z_2 \leqslant \frac{1}{2\theta} \|\frac{\partial z}{\partial t}\|_0^2 + \frac{\theta}{2} \|W_N\|_0^2,$$

with

$$\|\frac{\partial z}{\partial t}\|_0^2 = \|(I-P_N)\frac{\partial u}{\partial t}\|_0^2 \leqslant C(1+N^2)^{1-\tau} \|\frac{\partial u}{\partial t}\|_{\tau-1}^2.$$

Finally, for z_3 we have

$$z_3 = (Tz, W_N) = \varepsilon(Az, W_N) + (L_C z, W_N)$$

with

$$(Az, W_N) = \Big(b\frac{\partial z}{\partial x}, \frac{\partial W_N}{\partial x}\Big) \leqslant \frac{1}{2\beta} \|b\frac{\partial z}{\partial x}\|_0^2 + \frac{\beta}{2} \|\frac{\partial W_N}{\partial x}\|_0^2,$$

and

$$(L_C z, W_N) \leqslant \frac{1}{2\theta} \|L_C z\|_0^2 + \frac{\theta}{2} \|W_N\|_0^2.$$

$$\|L_C z\|_0^2 \leqslant C(1+N^2)^{1-\tau} \|u(t)\|_\tau^2 ,$$

(11.12) $$\|b\frac{\partial z}{\partial x}\|_0^2 \leqslant C\|z\|_1^2 \leqslant C(1+N^2)^{1-\tau} \|u(t)\|_\tau^2,$$

therefore

$$Z_3 < C(\frac{\varepsilon}{2\beta} + \frac{1}{2\theta})(1+N^2)^{1-\tau} \|u(t)\|_\tau^2 + \frac{\varepsilon\beta}{2} \|\frac{\partial W_N}{\partial x}\|_0^2 + \frac{\theta}{2} \|W_N\|_0^2 .$$

Gathering the terms Z_1, Z_2, Z_3, we find that:

$$\frac{1}{2}\frac{d}{dt} \|W_N\|_0^2 < \frac{3\theta}{2} \|W_N\|_0^2 + C(1+N^2)^{1-\tau}(\|u(t)\|_\tau^2 + \|\frac{\partial u}{\partial t}\|_{\tau-1}^2).$$

Applying Gronwall´s Lemma 11.2, proven later we deduce that

$$\|W_N(t)\|_0^2 < e^{\frac{3\theta}{2}t} (1+N^2)^{-1}(\|u_0\|_\tau^2 + \int_0^t (\|u(s)\|_\tau^2 + \|\frac{\partial u}{\partial t}(s)\|_{\tau-1}^2)ds),$$

where we have used the estimate established in Theorem 9.1 namely

$$\|W_N(0)\|_0^2 < C(1+N^2)^{-\tau}\|u_0\|_\tau^2.$$

Theorem 11.1 shows that

$$\|u(s)\|_\tau^2 < C\|u_0\|_\tau^2$$

with a constant C independent of ε so we conclude that

$$\|\frac{\partial u}{\partial t}\|_{\tau-1} < \varepsilon\|Au\|_{\tau-1} + \|Lu\|_{\tau-1} < C(\varepsilon\|u_0\|_{\tau+1} + \|u_0\|_\tau).$$

Q.E.D.

Lemma 11.2 (Gronwall´s Lemma): <u>Suppose that a differentiable function y satisfies the inequality</u>

(11.13) $$y´(t) < \alpha y(t) + g(t),$$

then:

$$y(t) < y_0 e^{\alpha t} + \int_0^t g(s) e^{\alpha(t-s)} ds.$$

Proof: We may rewrite (11.13) in the form

$$\frac{d}{dt}\left(y(t)e^{-\alpha t}\right) < g(t)e^{-\alpha t},$$

so integrating betwen 0 and t yields

$$y(t) < e^{\alpha t}\left(y_0 + \int_0^t g(s)e^{-\alpha s} ds\right).$$

Remark 11.1: The result obtained in Theorem 11.2 is not as strong as that of Theorem 9.1. In Theorem 11.2, we require that $u_0 \in H_p^{\tau+1}(I)$ instead of $H_p^{\tau}(I)$ which was all that was needed for the earlier error estimate.

In fact, we have merely established that

$$\left\|\frac{\partial u_\epsilon}{\partial t}\right\|_{\tau-1}^2 < C\left(\|u_0\|_\tau^2 + \epsilon\|u_0\|_{\tau+1}^2\right),$$

where the constant C is independent of ϵ.

In order to obtain a result which is as strong as Theorem 9.1, it is necessary to eliminate the term $\epsilon\|u_0\|_{\tau+1}^2$ in the right-hand side above. Now this is possible (see following example) in the constant coefficient case though at the cost of introducing a term in $1/t^2$ which diverges in the vicinity of $t = 0$.

<u>Example 11.1:</u>   Consider the particular case where

$$A = - \frac{d^2}{dx^2} \qquad \text{and} \qquad L = \frac{\partial}{\partial x} \, ,$$

that is to say where u_ε is the solution of

i) $\dfrac{\partial u_\varepsilon}{\partial t} - \varepsilon \dfrac{\partial^2 u_\varepsilon}{\partial x^2} + \dfrac{\partial u_\varepsilon}{\partial x} = 0$

ii) $u_\varepsilon(t,-\pi) = u_\varepsilon(t,\pi)$

(11.14)   (periodicity)

iii) $\dfrac{\partial u_\varepsilon}{\partial x}(t,-\pi) = \dfrac{\partial u_\varepsilon}{\partial x}(t,\pi)$

iv) $u_\varepsilon(0,x) = g(x)$ (initial condition).

In this case we know explicitly the Fourier coefficients of u_ε;   if

$$u_\varepsilon(x,t) = \sum_{n \in \mathbb{Z}} \hat{u}_n(t) e^{inx},$$

we have then, referring to (11.14)(i)

$$\frac{d\hat{u}_n}{dt} + (\varepsilon n^2 + in)\hat{u}_n = 0$$

$$\hat{u}_n(0) = \hat{g}_n$$

so

$$\hat{u}_n(t) = e^{-(\varepsilon n^2 + in)t} \hat{g}_n \, .$$

It is easily verified that

$$
1. \qquad \| u_\varepsilon(t) \|_0^2 = \sum_n \left| e^{-(\varepsilon n^2 + in)t} \right|^2 |\hat{g}_n|^2
$$

$$
= \sum_n e^{-2\varepsilon n^2 t} |g_n|^2 < \| g \|_0^2
$$

and that

$$
2. \qquad \| u_\varepsilon(t) \|_s^2 = \sum_n (1+n^2)^s e^{-2\varepsilon n^2 t} |\hat{g}_n|^2 ,
$$

is bounded for all s and $t > 0$ (but with a constant dependent on ε), and that for given ε, $g \in L^2 \to u_\varepsilon(t) \in H^s$ for $t > 0$ and any s (regularizing effect). We can also establish that (Theorem 11.1)

$$
\| u_\varepsilon(t) \|_s^2 < \| g \|_s^2 = \sum_n (1+n^2)^s |\hat{g}_n|^2 .
$$

3. Consider

$$
\frac{\partial u_\varepsilon}{\partial t} = \sum_n (-(\varepsilon n^2 + in)) e^{-(\varepsilon n^2 + in)t} \, \hat{g}_n \, e^{inx} .
$$

We have

$$
\left\| \frac{\partial u_\varepsilon}{\partial t} \right\|_s^2 = \sum_n (1+n^2)^s \left| \varepsilon n^2 + in \right|^2 e^{-2\varepsilon n^2 t} |\hat{g}_n|^2
$$

$$
= \sum_n (1+n^2)^s (\varepsilon^2 n^4 + n^2) e^{-2\varepsilon n^2 t} |\hat{g}_n|^2 .
$$

As the function

$$
\Phi(y) = y^2 e^{-2yt}
$$

is bounded by

$$\Phi(\frac{1}{t}) = \frac{1}{(te)^2} \, ,$$

we have

$$\varepsilon^2 n^4 e^{-2\varepsilon n^2 t} < \frac{1}{(te)^2} \, ;$$

therefore

$$\|\frac{\partial u_\varepsilon}{\partial t}\|_s^2 < \sum_n (1+n^2)^s (\frac{1}{(te)^2} + n^2)|\hat{g}_n|^2 < \|g\|_{s+1}^2 + \frac{1}{(te)^2} \|g\|_s^2 \, ,$$

which illustrates Remark 11.1.

(In this example with constant coefficients, we may calculate directly $P_N u_\varepsilon(t)$ without having to solve the discrete problem with the methods described in section 10.)

Remark 11.2: If $\varepsilon > 0$ is fixed, the regularization observed in the preceding example (which generalizes to the case of nonconstant coefficients[1]) ensures that $u(t) \in H_P^s(I)$ for all s, and t > 0. The order of the error may not be $O(N^{-s})$ for all s as one would expect because of possible errors in the approximation to the initial solution if it is not regular.

Remark 11.3: Suppose that we have to solve the problem (11.1) in the interval $]0,\pi[$ with the Dirichlet boundary conditions; (11.1)(iii) is replaced by

$$u(t,0) = u(t,\pi) = 0, \qquad \text{for all } t > 0.$$

---

[1] See Taylor, [17].

We will show that we may convert this problem to the one posed in the interval $I =]-\pi, \pi[$ with periodic boundary conditions.

To do so we will use the fact that the derivative of an odd function is even and vice versa.

Suppose that the solution u, the initial solution u_0, and the coefficients a, b and e are, for the moment, only defined on the interval $[0, \pi]$.

We can extend u, u_0 and a to be __odd__ over all I, and b and e to be __even__; for $x < 0$, we let

$$u(x) = -u(-x), \quad u_0(x) = -u_0(-x), \quad a(x) = -a(-x)$$

$$b(x) = b(-x), \quad e(x) = e(-x).$$

In this fashion $\frac{\partial u}{\partial x}$ will be even as will, $b \frac{\partial u}{\partial x}$ while, $\frac{\partial}{\partial x} b \frac{\partial u}{\partial x}$ and Au will be odd.

Similarly, $a \frac{\partial u}{\partial x}$ will be odd, au will be even, therefore $\frac{\partial}{\partial x} (au)$ is odd and Lu will thus be odd.

If the equation (11.1)(i) holds over $]0, \pi[$, it will hold also on the interval $]-\pi, 0[$ and at 0 (since an odd function is zero at the origin). On the other hand u is periodic. By the uniqueness of the solution of the periodic problem, we are brought back to solving a problem with periodic boundary conditions. However, even if the given initial u_0 and the coefficient a are regular for the problem with the Dirichlet boundary conditions on $]0, \pi[$, that is not necessarily so for the problem with periodic boundary conditions except if u_0 and a (at the same time their even order derivatives) vanish at 0 and π.

The Fourier method produces in fact an approximation to the function u on the interval $]0,\pi[$ by a sine series, an approximation which suffers from the same defects; we can only approximate well functions which along with all of their even order derivatives vanish at 0 and π.

We can also consider the problem with homogeneous Neumann boundary conditions.

$$\frac{\partial u}{\partial x}(t,0) = \frac{\partial u}{\partial x}(t,\pi) = 0;$$

in this case u and u_0 are extended over the entire interval as even functions, and the Fourier method will correspond to an approximation by a cosine series.

Remark 11.4:   A Nonhomogeneous equation.

Suppose that we have the problem

$$\frac{\partial u}{\partial t} + Tu = f$$

with $f \neq 0$ ((11.1)(i) and (ii)) being unchanged.   The discrete problem (11.8)(i) is replaced by

$$\frac{\partial u_C}{\partial t} + T_C u_C = f_C \qquad \text{with} \qquad f_C = P_C f.$$

The equivalent of equation (11.9) occuring in the proof of Theorem 11.2 is

$$\frac{\partial W_N}{\partial t} + T_C W_N = (T_C - T)\tilde{u}_N + \frac{\partial z}{\partial t} + Tz + f - f_C,$$

and there is a supplementary term to estimate, which depends on the regularity of f. (Note that the estimates given in Theorems 9.1 and 11.1 are always valid.)

12. The Solution of an Elliptic Problem

To conclude our study of the applications of Fourier series, we will now examine elliptic problems.

We consider the following stationary problem; find $u = u(x)$ such that

i) $Au = f,$ $x \in I,$

(12.1)

ii) $u(-\pi) = u(\pi),\ u'(-\pi) = u'(\pi)$ (periodic boundary conditions).

We suppose that the scalar γ introduced in the hypothesis (11.3) is negative so that (see (11.4))

(12.2) $(Au,u) \geq \alpha \|u\|_1^2$

with $\alpha = \min(\beta, -\gamma) > 0.$

The inequality (12.2) expresses the fact that the operator A is uniformly strongly elliptic on the space $H_P^1(I)$.

The Lax–Milgram lemma along with the regularity results for the elliptic problems (see Lions–Magenes, [11]) permits us to affirm the existence of a solution $u \in H_P^{s+2}(I)$ if $f \in H_P^s(I)$, for $s \geqslant 0$.

The discrete problem may be written naturally in the form

$$A_C u_C = f_C,$$

where A_C is defined in (11.6), and $f_C = P_N f$.

The operator A_C satisfies an inequality of uniform ellipticity (see Lemma 11.1):

$$(A_C u, u) \geqslant \alpha \|u\|_1^2, \qquad \text{for all } u \in S_N.$$

This will be useful in proving the following theorem.

Theorem 12.1: Let $\tau > 1$ be given; there exists a constant C such that if $f \in H_P^{\tau-2}(I)$ (and $u \in H_P^\tau(I)$), we have the (optimal) error estimate

$$\|u - u_C\|_1 \leqslant C(1+N^2)^{\frac{1-\tau}{2}} \|u\|_\tau.$$

Proof: We have, by setting $\tilde{u}_N = P_N u$ and $z = \tilde{u}_N - u$,

$$A_C \tilde{u}_N = (A_C - A)\tilde{u}_N + Az + f,$$

so for $W_N = \tilde{u}_N - u_C$,

$$A_C W_N = (A_C-A)\tilde{u}_N + Az + f - f_C,$$

and

$$\alpha\|W_N\|_1^2 < (A_C W_N, W_N) = ((A_C-A)\tilde{u}_N, W_N) + (Az, W_N) + (f-f_C, W_N).$$

Now, we have (see (11.6))

$$\left((A_C-A)\tilde{u}_N, W_n\right) = \left((P_C-I)\left(b\,\frac{\partial\tilde{u}_N}{\partial x}\right) + ((P_C-I)(e\tilde{u}_N), W_N\right)$$

$$< C(1+N^2)^{\frac{1-\tau}{2}}\,\|u\|_\tau\,\|W_N\|_1$$

and

$$(Az, W_N) = \left(b\,\frac{\partial z}{\partial x}\,,\,\frac{\partial W_N}{\partial x}\right) + (ez, W_N)$$

$$< C\left(\|z\|_1\,\|W_N\|_1 + \|z\|_0\,\|W_N\|_0\right)$$

$$< C\left((1+N^2)^{\frac{1-\tau}{2}} + (1+N^2)^{-\tau/2}\right)\|u\|_\tau\,\|W_N\|_1.$$

Finally,

$$(f-f_C, W_N) < \|f-f_C\|_{-1}\,\|W_N\|_1 < (1+N^2)^{\frac{1-\tau}{2}}\|f\|_{\tau-2}\,\|W_N\|_1,$$

where we have used Theorem 6.1 (with $s = -1$); then, noting that

$$\|u\|_\tau < C\|f\|_{\tau-2}\,, \qquad\qquad\qquad \text{(regularity result)},$$

we have

$$\alpha \|W_N\|_1^2 \leq C(1+N^2)^{\frac{1-\tau}{2}} \|f\|_{\tau-2} \|W_N\|_1,$$

and the result follows with

$$\|u-u_C\|_1 \leq \|u-\tilde{u}_N\|_1 + \|W_N\|_1.$$

Q.E.D.

Remark 12.1: If we choose $f_C = P_C f$ (the interpolant of f), then we must at least choose f in $H_P^1(I)$ for $P_C f$ to have a meaning. On the other hand, we only know in this case that

$$\|f-f_C\|_{-1} \leq C \|f-f_C\|_0,$$

which yields the nonoptimal error estimate

$$\|u-u_C\|_1 \leq C(1+N^2)^{\frac{1-\tau}{2}} \|u\|_{\tau+1}, \qquad \text{(for } \tau \geq 3 \text{)}.$$

1. A Review of Orthogonal Polynomials

Suppose $I =]a,b[$ is a given interval (bounded or not). Let $\omega : I \to \mathbb{R}^+$ be a weight function which is positive and continuous (and strictly positive on I).

We denote by $L^2_\omega(I)$ the space of measurable functions v from I into $\mathbb{C}$ such that

$$\| v \|_\omega \equiv \left(\int_I |v(x)|^2 \omega(x) dx \right)^{1/2} < +\infty.$$

$L^2_\omega(I)$ is a Hilbert space for the scalar product

$$(u,v)_\omega = \int_I u(x)\, \overline{v(x)}\, \omega(x) dx.$$

We will assume that

$$(1.1) \qquad \int_I x^n \omega \, dx < +\infty\, ; \qquad \text{for all } n \in \mathbb{N}$$

so that space $L^2_\omega(I)$ contains all the polynomials.

By othogonalization of the family of monomials

$$\{1, x, x^2, \cdots \},$$

we can obtain an orthonormal family of polynomials $(P_n)_{n \in \mathbb{N}}$ such that

i) $p_n \epsilon \; \mathbb{P}_n$ (space of polynomials of degree $<$ n)

(1.2)   ii)   the coefficient of x^n in p_n is strictly positive.

iii) $(p_n, p_m)_\omega = \delta_{nm}$ (orthonormality).

It is well known (cf. e.g., Davis-Rabinowitz [7]) that the polynomials p_n satisfy a recurrence relation of the following type

(1.3) $$x p_n = \alpha_n p_{n+1} + \beta_n p_n + \gamma_n p_{n-1}, \quad n > 1,$$

where $\alpha_n > 0$.   It is also well known that the zeros of p_n separate the zeros of p_{n+1}, and that the polynomial p_n has n distinct roots on I.

In particular (see (1.2)(ii)) this yields

i) $p_n(b) > 0$, $n \; \epsilon \; \mathbb{N}$

(1.4)

ii) $p_n(a) p_{n+1}(a) < 0$, $n \; \epsilon \; \mathbb{N}$

Example 1.1:   Chebyshev Polynomials.   In this case $\omega = (1-x^2)^{-1/2}$, and $I =]-1, +1[$. The Chebyshev polynomials are defined by $t_n(\cos \theta) = \cos n\theta$.

We now show that the t_n satisfy the recurrence relation

(1.5) $$2x t_n = t_{n+1} + t_{n-1}.$$

As

$$(1.6) \qquad \int_{-1}^{+1} f(x)\omega(x)dx = \int_{0}^{\pi} f(\cos\theta)d\theta,$$

we infer that

$$(t_n, t_m)_\omega = \int_{-1}^{+1} t_n(x)t_m(x)\omega(x)dx = \int_{0}^{\pi} \cos n\theta \cos m\theta \, d\theta,$$

whence

$$(t_n, t_m)_\omega = \begin{cases} \pi & \text{if } n = m = 0 \\ \frac{\pi}{2} & \text{if } n = m \neq 0 \ . \\ 0 & \text{if } n \neq m. \end{cases}$$

Therefore the family $(t_n)_{n \in \mathbb{N}}$ is orthogonal, but not orthonormal. We then set

$$p_n = \sqrt{\frac{2}{\pi}} \, t_n \qquad \text{for } n \geqslant 1$$

$$(1.7)$$

$$p_0 = \frac{1}{\sqrt{\pi}} \, t_0 = \frac{1}{\sqrt{\pi}} \qquad \text{for } n = 0.$$

Thus the recurrence relation (1.5) follows as $\alpha_n = \gamma_n = 1/2$, $\beta_n = 0$ for $n \geqslant 2$.

We note that the change of variable $x = \cos\theta$ transforms

$$u \in L_\omega^2(I) \quad \text{to} \quad \tilde{u} \in L^2(0,\pi),$$

by the formula $\tilde{u}(\theta) = u(\cos\theta)$.

This transformation is itself isometric since according to (1.6)

(1.8)
$$\int_0^\pi |\tilde{u}(\theta)|^2 \, d\theta = \int_I |u(x)|^2 \, \omega(x) dx.$$

For other examples of orthogonal polynomials (Legendre, Jacobi, Hermite Laguerre polynomials) we refer the reader to Davis and Rabinowitz [7].

2. An Introduction to the Numerical Integration Formulae of Gauss, Gauss-Radau and Gauss-Lobatto

We return to the general case of an interval I bounded or not with an arbitrary weight function ω. We denote by $(x_j)_{1 < j < N}$ the roots of the orthogonal polynomial p_N (of degree N).

We may choose some coefficients $(w_j)_{1 < j < N}$ such that the numerical integration formula

(2.1)
$$\int_I f(x)\omega(x) dx = \sum_{j=1}^N w_j f(x_j)$$

is __exact__ for $f \, \varepsilon \, \mathbb{P}_{N-1}$ (the w_j are the solutions of the linear system

$$\sum_{j=1}^N (x_j)^k \, w_j = \int_I x^k \, \omega dx, \qquad 0 < k < N-1,$$

whose matrix is invertible since the x_j are all distinct; it is the Van Der Monde matrix).

We recall that as the x_j are the roots of the orthogonal polynomial of order N, the formula (2.1) is in fact exact for $f \, \varepsilon \, \mathbb{P}_{2N-1}$; it is called the N point __Gauss formula__.

The <u>Gauss-Radau</u> formula, is defined in terms of the (N+1) roots of the polynomial q defined by

$$q(x) = P_N(a)P_{N+1}(x) - P_{N+1}(a)P_N(x),$$

which vanishes at x = a.

Let ξ_0 = a and $(\xi_j)_{1 \leq j \leq N}$ be the roots of polynomial q; we determine then (N+1) coefficients $(\omega_j)_{0 \leq j \leq N}$ such that the formula

(2.2)
$$\int_I f\omega dx \simeq \sum_{j=0}^{N} \omega_j f(\xi_j)$$

is exact for f ϵ $\mathbb{P}_N$.

The formula (2.2) is actually exact for f ϵ $\mathbb{P}_{2N}$. It is called the (N+1) point <u>Gauss-Radau formula</u> (associated with point a). There is, in fact, another Gauss-Radau formula associated with point b, obtained by replacing a by b in the definition of q. Finally, to obtain the (N+1) point <u>Gauss-Lobatto</u> formula, we use the N+1 roots of the polynomial $\tilde{q}$ defined by

$$\tilde{q}(x) = P_{N+1}(x) + \alpha P_N(x) + \beta P_{N-1}(x),$$

where α and β are determined by the condition test

$$\tilde{q}(a) = \tilde{q}(b) = 0.$$

Let $\tilde{\xi}_0$ = a, $(\tilde{\xi}_j)_{1 \leq j \leq N-1}$, $\tilde{\xi}_N$ = b be the roots of $\tilde{q}$.

In the usual fashion we determine the (N+1) coefficients $(\tilde{\omega}_j)_{0 \leq j \leq N}$, by demanding that the formula

(2.3)
$$\int_I f\omega dx \simeq \sum_{j=0}^{N} \tilde{\omega}_j f(\tilde{\xi}_j)$$

be exact for $f \in \mathbb{P}_N$. This formula is actually exact for $f \in \mathbb{P}_{2N-1}$ and is called the (N+1) point <u>Gauss-Lobatto formula</u>.

<u>Example 2.1</u>:   <u>The Case of the Chebyshev weight</u> $\omega(x) = (1-x^2)^{-1/2}$.

We are going to make explicit the Gauss, Gauss-Radau and Gauss-Lobatto formulae for the case where the weight ω is given by

$$\omega(x) = (1-x^2)^{-1/2}.$$

The corresponding formulae will be called the Gauss-Chebyshev, the Gauss-Radau-Chebyshev, and the Gauss-Lobatto-Chebyshev formulae. Let us begin with the Gauss-Radau-Chebyshev formula.

In Part A (see section 8) we have seen that the numerical integration formula

$$\frac{1}{2\pi} \int_{-\pi}^{\pi} g(\theta)d\theta \simeq \frac{1}{2N+1} \sum_{|j| \leqslant N} g(\theta_j),$$

where $\theta_j = j \frac{\pi}{N+1/2}$ is exact for $g \in S_{2N}$.

Limiting attention to functions g which are even in θ (that is to say the linear combinations of $(\cos n\theta)_{0 \leqslant n \leqslant 2N}$), we infer that the formula

$$\int_{0}^{\pi} g(\theta)d\theta \simeq \frac{\pi}{2N+1} (g(\theta_0) + 2 \sum_{j=1}^{N} g(\theta_j)),$$

is exact for all g of this form.

By the change of variable $x = \cos\theta$ (see (1.6)), we deduce that

$$(2.4) \qquad \int_{-1}^{+1} f(x)\omega(x)dx \simeq \frac{\pi}{2N+1}\left(f(\xi_0) + 2\sum_{j=1}^{N} f(\xi_j)\right),$$

with $\xi_j = \cos\theta_j$ is exact for $f \in \mathbb{P}_{2N}$.

This is therefore the <u>Gauss-Radau-Chebyshev formula</u> (associated with the point $+1$).

(The Gauss-Radau-Chebyshev formula associated with the point $x = -1$, would be obtained with $\xi_j = -\cos\theta_j$).

To obtain the <u>Gauss-Lobatto-Chebyshev</u> formula, we start with the numerical integration formula

$$\frac{1}{2\pi}\int_{-\pi}^{\pi} g(\theta)d\theta \simeq \frac{1}{2N}\sum_{j=-N+1}^{N} g(\widetilde{\theta}_j)$$

with $\widetilde{\theta}_j = j\frac{\pi}{N}$, which is exact for $g \in S_{2N-1}$ (see Part A, Remark 8.3).

Analogously to the above, we deduce that the formula

$$(2.5) \qquad \int_{-1}^{+1} f(x)\omega(x)dx \simeq \frac{\pi}{2N}\left(f(\widetilde{\xi}_0) + 2\sum_{j=1}^{N-1} f(\widetilde{\xi}_j) + f(\widetilde{\xi}_N)\right),$$

with $\widetilde{\xi}_j = \cos\widetilde{\theta}_j$, is exact for $f \in \mathbb{P}_{2N-1}$.

This is the Gauss-Lobatto-Chebyshev formula. Finally, to obtain the Gauss-Chebyshev formula, we remark that

$$\frac{1}{2\pi}\int_{-\pi}^{\pi} g(\theta)d\theta \simeq \frac{1}{2N}\sum_{\substack{|j'|=1/2}}^{N-1/2} g(\theta_{j'}),$$

where $\theta_{j'}$ is always given by $\widetilde{\theta}_{j'} = j'\frac{\pi}{N}$, but where index j' takes only fractional values ($j' + 1/2$ being a positive or negative integer), is exact for $g \in S_{2N-1}$.

In fact,

$$\frac{1}{2N} \sum_{|j^-|=1/2}^{N-1/2} e^{in\theta_{j^-}} = \begin{cases} 1 & \text{if } n = 0 \\ 0 & \text{if } 0 < |n| < 2N \\ -1 & \text{if } |n| = 2N \end{cases}.$$

As in the preceding case, we conclude that the formula

$$(2.6) \qquad \int_{-1}^{+1} f(x)\omega(x)dx \simeq \frac{\pi}{N} \sum_{j=1}^{N} f(x_j),$$

with $x_j = \cos(\frac{j-1/2}{N}\pi)$, is exact for $f \in \mathbb{P}_{2N-1}$; this is the Gauss-Chebyshev formula.

Example 2.2: The Jacobi Weight. We are going to make explicit the Gauss-Radau formula in the case of Jacobi weight $\omega(x) = \left(\frac{1+x}{1-x}\right)^{1/2}$. (This formula will be used in section 4.)

Let $g \in \mathbb{P}_{2N-1}$ be given, and $f = (1+x)g \in \mathbb{P}_{2N-1}$.

The Gauss-Lobatto-Chebyshev formula gives us

$$\int_{-1}^{+1} g(x)(1+x)(1-x^2)^{-1/2} dx = \frac{\pi}{2N} \left(2g(\widetilde{\xi}_o) + 2 \sum_{j=1}^{N-1} (1+\widetilde{\xi}_j)g(\widetilde{\xi}_j)\right).$$

(In fact, $\widetilde{\xi}_N = -1$, and thus $f(\widetilde{\xi}_N) = 0$).

We have thus shown that the formula

$$(2.7) \qquad \int_{-1}^{+1} g(x)\left(\frac{1+x}{1-x}\right)^{1/2} dx = \frac{\pi}{N} \left(g(1) + \sum_{j=1}^{N-1} (1+\widetilde{\xi}_j)g(\widetilde{\xi}_j)\right),$$

is exact for $g \in \mathbb{P}_{2N-2}$. This is the N point Gauss-Radau formula associated with the Jacobi weight

$$(\frac{1+x}{1-x})^{1/2} \quad \text{(and at the point } x = 1).$$

Similarly, we show that the formula

(2.8)
$$\int_{-1}^{+1} g(x)(\frac{1-x}{1+x})^{1/2} dx \simeq \frac{\pi}{N} (\sum_{j=1}^{N-1} (1-\tilde{\xi}_j)g(\tilde{\xi}_j) + g(-1)),$$

is exact for $g \in \mathbb{P}_{2N-2}$. This is thus a Gauss-Radau formula but associated with the weight

$$(\frac{1-x}{1+x})^{1/2} \quad \text{(and with the point } x = -1).$$

We are now going to concentrate our efforts on the Chebyshev weight $\omega(x) = (1-x^2)^{-1/2}$.

For this weight, the points of numerical integration for the Gauss-Lobatto and Gauss-Radau formulae, are the projections on the real axis of the M roots of unity of order M, where M is an even or odd integer.

This nice property will enable us to use the Fast Fourier Transform to compute the polynomial interpolant of a given function.

Before going to this, we shall first review some results obtained by Canuto-Quarteroni [4] about the best polynomial approximations of a function when using the Chebyshev weight.

3. The Approximation of a Function by Chebyshev Polynomials

We restrict ourselves in this section to the case where the weight ω is the Chebyshev weight

$$\omega(x) = (1-x^2)^{-1/2}.$$

We shall frequently use the mapping

$$L_\omega^2(I) \to L^2(-\pi,\pi)$$

(3.1)

$$u(x) \to \tilde{u}(\theta) \equiv u(\cos\theta).$$

From (1.7), we have

(3.2)
$$\|\tilde{u}\|_{L^2(-\pi,\pi)} = 2\|u\|_\omega .$$

The above mapping is therefore continuous and one to one.

Proposition 3.1: Let $P_N : L_\omega^2(I) \to \mathbb{P}_N$ be the orthogonal projection on the subspace $\mathbb{P}_N$ of polynomials of degree $\leq N$. For all $u \in L_\omega^2(I)$, we have

$$\|u - P_N u\|_\omega \to 0, \qquad\qquad \text{as } N \to \infty.$$

Proof: Suppose $\tilde{u}(\theta) = u(\cos\theta)$. From Part A (see Section 3) we may expand $\tilde{u}$ in Fourier series

(3.3)
$$\tilde{u}(\theta) = \sum_{n\in\mathbb{Z}} \hat{u}_n e^{in\theta},$$

In particular, let

$$(3.4) \qquad \tilde{u}_N(\theta) = \sum_{|n| < N} \hat{u}_n e^{in\theta} ,$$

then (see Section 3),

$$(3.5) \qquad \|\tilde{u} - \tilde{u}_N\|_{L^2(-\pi,\pi)} \to 0,$$

when $N \to \infty$.

As (by definition) $\tilde{u}$ is even in θ, we have

$$\hat{u}_n = \hat{u}_{-n} ,$$

and (from (3.3) and the definition of Chebyshev polynomials)

$$\tilde{u}(\theta) = \hat{u}_0 + 2 \sum_{n \geqslant 1} \hat{u}_n \cos n\theta$$

$$u(x) = \hat{u}_0 + 2 \sum_{n \geqslant 1} \hat{u}_n t_n(x).$$

Thus, from (3.4) we have

$$u_N(x) = \hat{u}_0 + 2 \sum_{n=1}^{N} \hat{u}_n t_n(x),$$

therefore $u_N = P_N u$ since the Chebyshev polynomials are orthogonal for the scalar product $(\cdot,\cdot)_\omega$ (see Section 1).

Finally, with (3.5) and (3.2) we have

$$\| u - P_N u \|_\omega \equiv \| u - u_N \|_\omega = \frac{1}{2} \| \tilde{u} - \tilde{u}_N \|_{L^2(-\pi,\pi)} \to 0, \quad \text{as} \quad N \to \infty.$$

Q. E. D.

We are now going to establish the error estimate for the quantity $\| u - P_N u \|_\omega$.

To do that, we introduce the family of weighted <u>Sobolev spaces</u>

$$H_\omega^m(I) = \left\{ u : u^{(\alpha)} \in L_\omega^2(I), \quad \alpha = 0, \cdots, m \right\}.$$

where $u^{(\alpha)}$ denotes the derivative of order α of u <u>in the distribution sense</u>. The space $H_\omega^m(I)$ is a Hilbert space for the norm

$$\| u \|_{m,\omega} \equiv \left(\sum_{\alpha=0}^{m} \| u^{(\alpha)} \|_\omega^2 \right)^{1/2}.$$

We will use the following result.

Theorem 3.1: <u>The mapping</u> $u \to \tilde{u}$ <u>defined by</u> $\tilde{u}(\theta) = u(\cos\theta)$ <u>for</u> $0 \le \theta \le \pi$ <u>is continuous from</u> $H_\omega^m(I)$ <u>to</u> $H^m(0,\pi)$.

<u>Proof:</u> In general, if we make a change of variable $x \to \theta$, where $x = \Phi(\theta)$, and Φ is a C^∞ function. Let

$$\tilde{u}(\theta) = u\big(\Phi(\theta)\big),$$

then

$$\left|\tilde{u}^{(m)}(\theta)\right| < C \sum_{\alpha=0}^{m} \left|u^{(\alpha)}(\Phi(\theta))\right|,$$

where C is a positive constant depending only on Φ and m. We deduce in the particular case where $\Phi(\theta) = \cos\theta$, that

$$\int_0^\pi \left|u^{(m)}(\theta)\right|^2 d\theta < C \sum_{\alpha=0}^{m} \int_I \left|u^{(\alpha)}(x)\right|^2 \omega(x)dx,$$

and this yields the desired result.

Q. E. D.

Remark 3.1: The mapping $u \to \tilde{u}$ defined in Theorem 3.1 is an isomorphism from $L^2_\omega(I)$ onto $L^2(0,\pi)$ (and it is also isometric).

However, the image of $H^m_\omega(I)$ is not the space $H^m(0,\pi)$ for $m \neq 0$, but a smaller space.

For example, the space

$$X = \left\{u : (1-x^2)^{1/4} u' \in L^2(I), (1-x^2)^{-1/4} u \in L^2(I)\right\}$$

(which contains strictly the space:

$$H^1_\omega(I) = \left\{u : (1-x^2)^{-1/4} u' \in L^2(I), (1-x^2)^{-1/4} u \in L^2(I)\right\})$$

has for its image $H^1(0,\pi)$.

In fact, the change of variable formula

$$\int_I \left(\overline{\sqrt{1-x^2}} |u'|^2 + \frac{1}{\sqrt{1-x^2}} |u|^2\right)dx = \int_0^\pi \left(|\tilde{u}'|^2 + |\tilde{u}|^2\right)d\theta,$$

shows that the mapping $u \to \tilde{u}$ is an isometry from X onto $H^1(0,\pi)$.

Theorem 3.2: The mapping $u \to \tilde{u}$ defined by $\tilde{u}(\theta) \equiv u(\cos\theta)$ for $-\pi < \theta < \pi$, is continuous from $H_\omega^m(I)$ into $H_P^m(-\pi,\pi)$ (where $H_P^m(-\pi,\pi)$ is the periodic Sobolev space of order m defined in Part A, Section 6).

Proof: The function $\tilde{u}$ defined by $\tilde{u}(\theta) \equiv u(\cos\theta)$ is obviously even; consequently, its derivatives of even order are even, and its derivatives of odd order are odd. According to Theorem 3.1, if $u \in H_\omega^m(I)$, then the restriction of $\tilde{u}$ to $]0,\pi[$ is in $H^m(0,\pi)$; likewise, its restriction to $]-\pi,0[$ is in $H^m(-\pi,0)$.

In order that $\tilde{u} \in H_P^m(-\pi,\pi)$, it suffices that all its derivatives of order less than or equal to $m-1$ be continuous and periodic (of period 2π); as $\tilde{u}$ is even, this is clearly true for its even order derivatives. It is necessary however to verify that its odd order derivatives ($\leq m-1$) vanish at $\theta = 0$ and $\theta = \pm\pi$. This follows from the fact that the derivatives of $\Phi(\theta) \equiv \cos\theta$ are zero for $\theta = 0$ and $\theta = \pm\pi$.

In fact, we have for example

$$\tilde{u}'(\theta) = -\sin\theta\, u'(\cos\theta)$$

$$\tilde{u}'''(\theta) = -\sin^3\theta\, u'''(\cos\theta) + 3\sin\theta\cos\theta\, u''(\cos\theta) + \sin\theta\, u'(\cos\theta),$$

which shows that if $u \in H_\omega^m(I)$ with $m \geq 2$, then

$$\tilde{u}'(0) = \tilde{u}'(\pm\pi) = 0,$$

and that if $u \in H_\omega^m(I)$ with $m \geqslant 4$, we have in addition

$$\tilde{u}'''(0) = \tilde{u}'''(\pm\pi) = 0,$$

and the result follows for $0 \leqslant m \leqslant 5$.

The proof of the general case is left to the reader.

Q. E. D

Remark 3.2: We refer the reader to Canuto-Quarteroni [4] for the definition of spaces $H_\omega^s(I)$, s noninteger, and for a generalization of the preceding theorems to the case where the integer m is replaced by positive real s.

Theorem 3.3: Let s > 0 be given. There exists a constant C such that

$$\|u - P_N u\|_\omega \leqslant CN^{-s} \|u\|_{s,\omega}$$

for all $u \in H_\omega^s(I)$.

Proof: Let $u_N \equiv P_N u, \tilde{u}_N(\theta) \equiv u_N(\cos\theta)$ and $\tilde{u}(\theta) = u(\cos\theta)$. From (3.2), we have

$$\|u - P_N u\|_\omega = \|u - u_N\|_\omega = \frac{1}{2} \|\tilde{u} - \tilde{u}_N\|_{L^2(-\pi,\pi)}.$$

Now, (see the proof of Proposition 3.1), $\tilde{u}_N$ happens to be equal to the Fourier series of $\tilde{u}$ truncated to order N.

According to Theorem 6.1 of Part A, we have therefore

$$\|\tilde{u}-\tilde{u}_N\|_{L^2(-\pi,\pi)} \leqslant C \, N^{-s} \, \|\tilde{u}\|_{H^s_P(-\pi,\pi)} \, .$$

On the other hand, Theorem 3.2[1] yields

$$\|\tilde{u}\|_{H^s_P(-\pi,\pi)} \leqslant C \, \|u\|_{H^s_\omega(I)} \, ,$$

which proves the result.

Q. E. D.

We will now establish an estimate for $\|u-P_N u\|_{\sigma,\omega}$ which is the error between u and its projection on the subspace $\mathbb{R}_N$ in the norm of $H^\sigma_\omega(I)$.

We introduce the following convention; if $(b_n)_{n \, \epsilon \, \mathbb{N}}$ denotes any sequence, we denote by

$$\sum_{\ell=m}^{n}{}' \, b_\ell \overset{def}{=} \sum_{\ell'=0}^{[\frac{n-m}{2}]} b_{m+2\ell'} \, ,$$

where $[\alpha]$ denotes the integer part of any real number α.

We define also the sequence $(c_k)_{k \epsilon \, \mathbb{N}}$ by

$$c_k = \begin{cases} 2 & \text{if } k = 0 \\ 1 & \text{otherwise} \end{cases} \, ,$$

---

[1] in the case of nonintegers s, see the Remark 3.2.

this will simplify the presentation of results.

Lemma 3.1: <u>Let</u> $u \in \mathbb{P}_N$ <u>be a polynomial of degree</u> N <u>and</u>

$$u = \sum_{k=0}^{N} a_k \, t_k$$

<u>be its expansion in Chebyshev polynomials.</u> <u>Then its derivative</u> u' <u>is given</u> <u>by</u>

$$u' = \sum_{k=0}^{N} b_k \, t_k$$

where

$$b_k = \frac{2}{c_k} \sum_{\ell=k+1}^{N} {}' \, \ell a_\ell \quad .$$

Proof: The following formulae are easily confirmed

$$t_0 = t_1'$$

$$t_n = \frac{1}{2} \left(\frac{t_{n+1}'}{n+1} - \frac{t_{n-1}'}{n-1} \right) , \qquad \text{for } n > 1.$$

We have then

$$u' = \sum_{k=0}^{N} b_k t_k = b_0 t_1' + \frac{1}{2} \sum_{k=1}^{N} b_k \left(\frac{t_{k+1}'}{k+1} - \frac{t_{k-1}'}{k-1} \right).$$

Thus

$$u' = \sum_{k=0}^{N} a_k t'_k \quad .$$

The following formulae follow

$$b_0 - \frac{b_2}{2} = a_1$$

$$\frac{1}{2n} \left(b_{n-1} - b_{n+1} \right) = a_n, \qquad 2 < n < N-2$$

$$\frac{b_{N-2}}{2(N-1)} = a_{N-1}$$

$$\frac{b_{N-1}}{2N} = a_{N-2} ,$$

whence the result, solving this system of equations (upper triangular matrix) by substitution.

Q. E. D.

Lemma 3.2 (Inverse Inequality): <u>Let</u> $s > 0$ <u>be given.</u> <u>There exists a constant</u> C <u>such that:</u>

(3.6) $$\| u \|_{s,\omega} < CN^{2s} \| u \|_{\omega}, \qquad \underline{\text{for all}} \ u \in \mathbb{R}_N,$$

<u>and for all</u> $N > 0$.

<u>Proof:</u> Let us begin by establishing the result for $s = 1$. Let $u \in \mathbb{R}_N$; we have

$$u = \sum_{k=0}^{N} a_k t_k$$

and

$$u^{\prime} = \sum_{k=0}^{N} b_k t_k ,$$

with

$$b_k = \frac{2}{c_k} \sum_{\ell=k+1}^{N} {}^{\prime} \ell a_\ell ,$$

from Lemma 3.1. Noting that:

$$(t_n, t_m)_\omega = c_n \frac{\pi}{2} \delta_{nm} ,$$

(see Section 1, example 1.1) we obtain

$$\| u^{\prime} \|_\omega^2 = \frac{\pi}{2} \sum_{k=0}^{N} c_k \, |b_k|^2 = \pi \sum_{k=0}^{N} \frac{2}{c_k} \left| \sum_{\ell=k+1}^{N} {}^{\prime} \ell a_\ell \right|^2 .$$

Now, the Schwarz inequality yields

$$\left| \sum_{\ell=k+1}^{N} \ell a_\ell \right|^2 < \left(\sum_{\ell=k+1}^{N} \ell^2 \right) \left(\sum_{\ell=k+1}^{N} |a_\ell|^2 \right) < N^3 \sum_{\ell=0}^{N} |a_\ell|^2 < CN^3 \, \| u \|_\omega^2 .$$

We deduce

$$\| u^{\prime} \|_\omega^2 < CN^4 \, \| u \|_\omega^2$$

whence the result for $s = 1$.

A repeated application of this theorem furnishes the result for any positive integer s.

We refer the reader to Canuto-Quarteroni [4] for a proof in the case of noninteger s.

Q. E. D.

Remark 3.2: In inequality (3.6) the exponent of N is optimal (although worse than that obtained in the case of Fourier series, see Part A, Proposition 8.1). In fact, (see Canuto-Quarteroni [4], we may find polynomials of degree N such that

$$\frac{\| P_N \|_{m,\omega}}{\| P_N \|_{\omega}} \sim N^{2m} .$$

We present the following result (prove in Canuto-Quarteroni, [4]) which constitutes an extension of Lemma 3.1.

Proposition 3.2: Let u be a sufficiently regular function such that

$$u = \sum_{k=0}^{\infty} a_k t_k ,$$

then we have

$$u´ = \sum_{k=0}^{\infty} b_k t_k$$

with

$$b_k = \frac{2}{c_k} \sum_{\ell=k+1}^{\infty} {}^{´} \ell a_\ell .$$

In order to estimate the error between u and $P_N u$ in the space $H_\omega^1(I)$, it is necessary to estimate

$$\| u´ - (P_N u)´ \|_\omega .$$

Now, contrary to the case of Fourier series (Part A), here $(P_N u)^\checkmark$ and $P_N u^\checkmark$ are not identical (note that $(P_N u)^\checkmark$ is a polynomial of degree $\leq N-1$, and $P_N u^\checkmark$ is a polynomial of degree N).

We then begin by estimating

$$\| P_N u^\checkmark - (P_N u)^\checkmark \|_\omega$$

Lemma 3.3:   <u>Suppose</u> $u \in H_\omega^s(I)$ <u>then we have the inequality</u>

$$\| P_N u^\checkmark - (P_N u)^\checkmark \|_\omega \leq CN^{-s+ 3/2} \| u \|_{s,\omega}.$$

<u>Proof:</u>   Let

$$q_N = P_N u^\checkmark - (P_N u)^\checkmark,$$

$$\tilde{u} = \sum_{k \geq 0} a_k t_k,$$

$$\tilde{u}^\checkmark = \sum_{k \geq 0} b_k t_k.$$

From Lemma 3.1, we have

$$b_k = \frac{2}{c_k} \sum_{\ell=k+1}^{\infty}{}^{\checkmark} \ell a_\ell.$$

Similarly as

$$P_N u = \sum_{k=0}^{N} a_k t_k,$$

we have

$$(P_N u)' = \sum_{k=0}^{N} b_k^N \, t_k$$

with

$$b_k^N = \frac{2}{c_k} \sum_{\ell=k+1}^{N} \ell a_\ell .$$

We deduce that

$$q_N = \sum_{k=0}^{N} \gamma_k \, t_k \, ,$$

with

$$\gamma_k = b_k - b_k^N = \frac{2}{c_k} \Big[\sum_{\ell'=0}^{\infty} (k+2\ell'+1)a_{k+2\ell'+1} - \sum_{\ell'=0}^{m'} (k+2\ell'+1)a_{k+2\ell'+1} \Big],$$

where

$$m' = \begin{cases} \dfrac{n-k-1}{2} & \text{if } N-k \text{ is odd} \\[2mm] \dfrac{n-k-2}{2} & \text{if } N-k \text{ is even} \end{cases} .$$

Therefore

$$c_k \gamma_k = 2 \sum_{\ell'=m'+1}^{\infty} (k+2\ell'+1)a_{k+2\ell'+1} .$$

That is to say

$$c_k \gamma_k = \begin{cases} 2 \displaystyle\sum_{\ell=N}^{\infty}{}' \ell a_\ell \equiv b_{N+1} & \text{if } N-k \text{ is odd} \\[4mm] 2 \displaystyle\sum_{\ell=N+1}^{\infty}{}' \ell a_\ell \equiv b_N & \text{if } N-k \text{ is even} \end{cases} .$$

We have then demonstrated that if N is even

$$q_N = \frac{1}{2} b_N t_0 + b_{N+1} t_1 + b_N t_2 + \cdots + b_N t_N \, ,$$

and if N is odd

$$q_N = \frac{1}{2} b_{N+1} t_0 + b_N t_1 + b_{N+1} t_2 + \cdots + b_N t_N,$$

that is to say

$$q_N = \begin{cases} b_N \phi_0^N + b_{N+1} \phi_1^N & \text{if } N \text{ even} \\ \\ b_N \phi_1^N + b_{N+1} \phi_0^N & \text{otherwise} \end{cases} \, ,$$

where

$$\phi_0^N = \sum_{\ell=0}^{N} {}^{\prime} \frac{1}{c_\ell} t_\ell , \quad \phi_1^N = \sum_{\ell=1}^{N} {}^{\prime} t_\ell .$$

As the functions ϕ_0^N and ϕ_1^N are orthogonal, we have

$$\| q_N \|_\omega^2 = \begin{cases} |b_N|^2 \, \| \phi_0^N \|_\omega^2 + |b_{N+1}|^2 \, \| \phi_1^N \|_\omega^2 & \text{if } N \text{ even} \\ \\ |b_N|^2 \, \| \phi_1^N \|_\omega^2 + |b_{N+1}|^2 \, \| \phi_0^N \|_\omega^2 & \text{if } N \text{ odd} \end{cases} \quad .$$

Now, from Theorem 3.3, we have

$$\| u^{\prime} - P_{N-1} u^{\prime} \|_\omega \leqslant C(N-1)^{1-s} \, \| u^{\prime} \|_{s-1} \leqslant C N^{1-s} \, \| u \|_s .$$

Since

$$u^{\prime} - P_{N-1} u^{\prime} = \sum_{n \geqslant N} b_n t_n ,$$

we have established that

$$|b_n| < CN^{1-s} \|u\|_s \qquad \text{for all } n \geqslant N.$$

Finally, as

$$\|\phi_0^N\|_\omega^2 \sim \|\phi_1^N\|_\omega^2 \sim N ,$$

we deduce that

$$\|q_N\|_\omega^2 < CN^{3-2s} \|u\|_s^2 .$$

Q. E. D.

Corollary 3.1: For all ρ and σ such that $0 \leqslant \rho \leqslant \sigma-1$, there exists a constant C such that

$$\|P_N u^{\cdot} - (P_N u)^{\cdot}\|_{\rho,\omega} < CN^{2\rho-\sigma+3/2} \|u\|_{\sigma,\omega}$$

for all $u \in H_\omega^\sigma(I)$.

(Apply Lemmas 3.2 and 3.3. Following Remark 3.2, this result may be extended to the case where ρ and σ are real.)

Theorem 3.4: For all μ and σ $0 \leqslant \mu \leqslant \sigma$, there exists a constant C such that

$$\|u - P_N u\|_{\mu,\omega} < CN^{e(\mu,\sigma)} \|u\|_{\sigma,\omega}$$

<u>for all</u> $u \in H_\omega^\sigma(I)$, <u>where</u>

$$e(\mu,\sigma) = \begin{cases} 2\mu - \sigma - \frac{1}{2} & \text{for } \mu \geq 1 \\ \frac{3}{2}\mu - \sigma & \text{for } 0 \leq \mu \leq 1 \end{cases} .$$

<u>Proof:</u>   (We restrict ourselves to the case of integer μ).   The result is obviously true for $\mu = 0$.   Suppose by induction that it is so for all $\mu = 0, \cdots, m-1$.   From the relation

$$\|v\|_{m,\omega}^2 = \|v^{(m)}\|_\omega^2 + \|v\|_{m-1,\omega}^2 \leq \|v'\|_{m-1,\omega}^2 + \|v\|_{m-1,\omega}^2 ,$$

which is true for all $v \in H_\omega^m(I)$, we get, using the induction hypothesis

$$\|u - P_N u\|_{m,\omega}^2 \leq \|u' - (P_N u)'\|_{m-1,\omega}^2 + CN^{2e(m-1,\sigma)} \|u\|_{\sigma,\omega}^2 .$$

Now using once again the induction hypothesis, and Lemma 3.3 we get

$$\|u' - (P_N u)'\|_{m-1,\omega} \leq \|u' - P_N u'\|_{m-1,\omega} + \|P_N u' - (P_N u)'\|_{m-1,\omega}$$

$$\leq CN^{e(m-1,\sigma-1)} \|u'\|_{\sigma-1,\omega} + CN^{2(m-1)-\sigma+3/2} \|u\|_{\sigma,\omega} ,$$

we deduce

$$\|u - P_N u\|_{m,\omega} \leq C\left[\left(N^{e(m-1,\sigma-1)} + N^{2(m-1)-\sigma+3/2}\right)^2 + N^{2e(m-1,\sigma)}\right]^{1/2} \|u\|_{\sigma,\omega}$$

$$\leqslant CN^{e(m,\sigma)} \|u\|_{\sigma,\omega} \ .$$

(In fact $e(m-1,\sigma-1)$ and $e(m-1,\sigma)$ are bounded by $e(m,\sigma)$; the dominant term is then the second term

$$N^{2(m-1)-\sigma+3/2} = N^{e(m,\sigma)} \qquad \text{for} \quad m > 1.)$$

Q. E. D.

Remark 3.4: The exponent N in the upper bound found for $\|u - P_N u\|_{\mu,\omega}$ cannot be improved; we refer to Canuto-Quarteroni [4], for counter examples.

4. Approximation by the Interpolation Operator

In the previous section, we have established error estimates for $u - P_N u$, where P_N is the projection operator of $L^2_\omega(I)$ on $\mathbb{P}_N$.

This result does not suffice in applications where boundary conditions must be taken into account.

As in the case of Fourier series (see Part A, Section 8) it is necessary to define an interpolation operator

$$P_C : C^0(I) \rightarrow \mathbb{P}_N$$

defined by

$$(P_C u)(y_j) = u(y_j) \qquad 0 \leqslant j \leqslant N,$$

where $(y_j)_{0 \leqslant j \leqslant N}$ are $(N+1)$ distinct points of the interval I. The operator P_C is thus defined in a unique fashion (Lagrange interpolation).

We consider first the case where

$$y_j = \xi_j \equiv \cos \frac{2j\pi}{2N+1} \; ,$$

that is to say, we use as interpolation points the <u>Gauss-Radau-Chebyshev</u> points (see Example 2.1). Introducing again the change of variable $x = \cos\theta$

$$u(x) \;\rightarrow\; \tilde{u}(\theta) \quad \text{with} \quad \tilde{u}(\theta) \equiv u(\cos\theta).$$

We note that the operator P_C is related to the interpolation operator $\tilde{P}_C : C_P^0(-\pi,\pi) \rightarrow S_N$ defined by

$$(\tilde{P}_C\tilde{u})(\theta_j) = u(\theta_j) \qquad\qquad |j| \leqslant N,$$

(with $\theta_j \equiv \dfrac{j\pi}{N+\frac{1}{2}}$) and which has been studied (under another name) in Part A (see Section 8).

More precisely, we have

$$\tilde{P}_C\tilde{u} = \widetilde{P_C u}.$$

Theorem 4.1: <u>Let $s > \frac{1}{2}$ and σ be given such that $0 \leqslant \sigma \leqslant s$. There exists a constant C such that</u>

$$\| u - P_C u \|_{\sigma,\omega} \;\leqslant\; C \, N^{2\sigma-s} \, \|u\|_{s,\omega}, \qquad \underline{\text{for all}} \quad u \in H_\omega^s(I).$$

Proof: Let us begin by establishing the result for $\sigma = 0$. Setting $\tilde{u}(\theta) = u(\cos \theta)$, we have (see Theorem 3.2)

$$\|\tilde{u}\|_{H_P^s(-\pi,\pi)} \leq C\|u\|_{s,\omega} .$$

From Part A (Theorem 9.1), we have for $s > \frac{1}{2}$

$$\|\tilde{u} - \tilde{P}_C \tilde{u}\|_{L^2(-\pi,\pi)} \leq C N^{-s} \|\tilde{u}\|_{H_P^s(-\pi,\pi)}$$

whence

$$(4.1) \qquad \|u - P_C u\|_{0,\omega} = \frac{1}{2} \|\tilde{u} - \tilde{P}_C \tilde{u}\|_{L^2(-\pi,\pi)} \leq C N^{-s} \|u\|_{s,\omega} .$$

For $\sigma > 0$, we note that, according to inverse inequality (Lemma 3.2)

$$\|u - P_C u\|_{\sigma,\omega} \leq \|u - P_N u\|_{\sigma,\omega} + C N^{2\sigma} \|P_N u - P_C u\|_{0,\omega} .$$

The conclusion follows from Theorem 3.4 and the inequality (4.1).

Q. E. D.

Remark 4.1: We note that the approximation properties of the operator P_C are weaker than those of P_N, at least when $\sigma > 0$. Actually, let $\|\cdot\|_\infty$ denote the norm $C^0(I)$ defined by

$$\|u\|_\infty = \max_{x \in I} |u(x)| ,$$

it is well known (see e.g., Rivlin [15]) that

$$\| u - P_C u \|_\infty \leq (1 + \Lambda_N) \| u - P_N u \|_\infty \ ,$$

where Λ_N is called the Lebesque constant.

Actually Brutman [3] has proved that Λ_N grows like $\log N$.

If the interpolation points were chosen in an arbitrary way the growth of the Lebesque constant Λ_N could be much worse. In fact for equally spaced points Λ_N grows exponentially fast. This is, of course, one good reason for not using equally spaced points, another reason being that the computation of $P_C u$ is ill-conditioned for such points.

Remark 4.2: Theorem 4.1 is established when the interpolation points y_j are those of Gauss-Radau-Chebyshev formula associated with the point $x = 1$.

We have an analogous result in the case where the interpolation points are those of Gauss-Radau-Chebyshev formula associated with the point $x = -1$ (change x to $-x$).

Let us consider now the case where the interpolation points are those of Gauss-Lobatto-Chebyshev formula.

$$y_j = \tilde{\xi}_j = \cos \frac{j\pi}{N} \ , \qquad j = 0, \cdots, N.$$

Suppose π_C is the interpolation operator $C^0(\bar{I}) \to \mathbb{P}_N$ (defined by $(\pi_C u)(\tilde{\xi}_j) = u(\tilde{\xi}_j)$), we have the following result.

Theorem 4.2: Let $s > \frac{1}{2}$ and σ be given such that $0 \leq \sigma \leq s$. There exists a constant C such that

$$\|u - \pi_C u\|_{\sigma,\omega} \leqslant C N^{2\sigma-s} \|u\|_{s,\omega}$$

<u>for all</u> $u \in H_\omega^s(I)$.

The proof of this result is in every respect analogous to the proof of Theorem 4.1 because the image of the operator π_C under the change of the variable $x \rightarrow \theta$ is an interpolation operator which has already been studied in Part A (see Remark 8.3 and formula (8.20)).

5. The Solution of the Advection Equation

We consider the advection equation in the interval $I =]-1,+1[$

$$\text{i)} \quad \frac{\partial u}{\partial t} + a(x) \frac{\partial u}{\partial x} = 0 \qquad , x \in I, t > 0.$$

(5.1)

$$\text{ii)} \quad u(-1,t) = g(t) \qquad , t > 0,$$

$$\text{iii)} \quad u(x,0) = u_0(x) \qquad , x \in I.$$

Unlike the problem studied in Part A (see Sections 7 and 9) the boundary conditions are not periodic.

We suppose that coefficient $a \in C^\infty(\overline{I})$ is strictly positive in $\overline{I}$.

We consider for simplicity the case of a homogeneous boundary condition $(g(t) \equiv 0)$.

We are going to approximate the problem (5.1) using a <u>collocation method</u> which we now describe. Let

$$U_N = \{p \; \varepsilon \; \mathbb{P}_N \; : \; p(-1) = 0\}.$$

and let $(x_j)_{j=1,\cdots,N}$ be N given points in the interval I.

The approximate problem will then be the following Find $u_N(t) \; \varepsilon \; \mathbb{P}_N$ such that

i) $(\dfrac{\partial u_N}{\partial t} + a \dfrac{\partial u_N}{\partial x}) \; (x_j) = 0$, $j = 1,\cdots,N,$ $t > 0.$

(5.2)   ii) $u_N(-1,t) = 0$, $t > 0$

iii) $u_N(x,0) = u_{0N}(x),$, $x \; \varepsilon \; I,$

where $u_{0N} \; \varepsilon \; U_N$ will be fixed subsequently.

The essential problem which is posed is the following

<u>How does one choose the collocation points</u> x_j <u>so that the method is</u> <u>stable?</u>   (In other words, so that the u_N of the system of ordinary differential equations will not grow exponentially.)

Numerical experimentation shows that the correct choice of the collocation points is crucial to the success of the method.

<u>Method A:</u>   (See Gottlieb [8].)

We first study the points

(5.3)
$$x_j = -\cos \frac{j\pi}{N+1} , \qquad j = 1, \cdots, N$$

(which are used <u>both</u> by the (N+2)-point Gauss-Lobatto-Chebyshev formula and by the (N+1)-point Gauss-Radau formula for weight

$$\omega_1 \equiv \left(\frac{1-x}{1+x}\right)^{1/2}$$

and associated with point $x = 1$, (see Section 2).

Theorem 5.1: <u>With the choice (5.3) for the collocation points, we have the stability for the discrete norm</u> $\|\cdot\|_N$ <u>associated with the discrete scalar product</u>

$$(u,v)_N = \sum_{j=0}^{N} \frac{\omega_j}{a(x_j)} u(x_j)v(x_j),$$

<u>where</u>

$$x_0 = -1, \quad \omega_0 = \frac{\pi}{N+1} \quad \underline{and} \quad \omega_j = (1-x_j) \frac{\pi}{N+1} .$$

<u>That is to say, we have</u>

$$\|u_N(t)\|_N^2 < \|u_N(0)\|_N^2 , \qquad \underline{for~all}~~t > 0.$$

Proof: According to (5.2), we have

(5.4)
$$\frac{\partial u_N}{\partial t} (x_j) + a(x_j) \frac{\partial u_N}{\partial x} (x_j) = 0, \qquad j = 1, \cdots, N.$$

We have seen (2.8) that the formula

(5.5)
$$\int_I g(x)\omega_1(x)dx \simeq \sum_{j=0}^N \omega_j \, g\,(x_j),$$

(where $\omega_1(x) = \left(\frac{1-x}{1+x}\right)^{1/2}$) was exact for $g \in \mathbb{P}_{2N}$ (this is a (N+1)-point Gauss–Radau formula.

Multiplying (5.4) by $\omega_j \dfrac{u_N(x_j)}{a(x_j)}$ and summing, we obtain (by noting that $u_N(x_0) = 0$ according to (5.2ii))

$$\sum_{j=0}^N \frac{\omega_j}{a(x_j)} u_N(x_j) \frac{\partial u_N}{\partial t}(x_j) + \sum_{j=0}^N \omega_j \, u_N(x_j) \frac{\partial u_N}{\partial x}(x_j) = 0,$$

that is, to say

(5.6)
$$\left(u_N, \frac{\partial u_N}{\partial t}\right)_N + \int_I u_N \frac{\partial u_N}{\partial x} \omega_1 \, dx = 0.$$

Now, integrating by parts (and noting that $u_N(-1) = 0$ and $\omega_1(1) = 0$)

$$\int_I u_N \frac{\partial u_N}{\partial x} \omega_1 \, dx = -\int_I u_N \frac{\partial}{\partial x}(\omega_1 u_N)dx$$

whence

$$2 \int_I u_N \frac{\partial u_N}{\partial x} \omega_1 \, dx = -\int_I u_N^2 \, \omega_1' \, dx \; > \; 0.$$

Returning to (5.6), we see that

$$\frac{1}{2} \frac{d}{dt} \|u_N(t)\|_N^2 = \left(u_N, \frac{\partial u_N}{\partial t}\right)_N < 0,$$

which proves the result.

Q.E.D.

Remark 5.1: Suppose α and C are such that $0 < \alpha \leq a(x) \leq C$. According to (5.5), we have

$$(5.7) \qquad \frac{1}{C} \|v_N\|_{\omega_1}^2 < \|v_N\|_N^2 \equiv \sum_{j=0}^{N} \frac{\omega_j}{a(x_j)} |v_N(x_j)|^2 < \frac{1}{\alpha} \|v_N\|_{\omega_1}^2 ,$$

for all $v_N \in \mathbb{R}_N$. Therefore, from Theorem 5.1 we get that for all $t > 0$

$$\|u_N(t)\|_{\omega_1}^2 < C\|u_N(t)\|_N^2 < +\infty,$$

which means stability in $L_{\omega_1}^2$.

We will show later that method A is easily implemented using Fast Fourier Transforms (see Section 7).

Remark 5.2: Choice of the Weight ω_1

Let us consider the particular case when the coefficient a is a constant. The exact solution of problem (5.1) is then

$$u(x,t) = u_0(x-at),$$

so that we may have

$$\|u(t)\|_{\omega_1} < \|u_0\|_{\omega_1} , \qquad \text{for } t > 0,$$

only if ω_1 is decreasing.

We note that this is what happens in Method A if

$$\omega_1 \equiv \left(\frac{1-x}{1+x}\right)^{1/2}.$$

This explains why we <u>cannot</u> have stability for the norm associated with the Chebyshev weight $\omega = (1-x^2)^{-1/2}$.

Theorem 5.2: <u>Suppose</u> $\sigma > 1/2$, $s > 2(1+\sigma)$ <u>and</u> $T > 0$ <u>are given. Then</u> <u>if</u> $u(t) \in H^s_\omega(I)$ <u>for</u> $0 < t < T$, <u>there exists</u> $C > 0$ <u>such that</u>

$$\|u(t) - u_N(t)\|_{\omega_1} < C \, N^{2(1-\sigma)-s} + \|u_{0N} - u_0\|_{\omega_1}$$

<u>for all</u> $t < T$.

Proof: Let $\xi_j = -\cos\frac{j\pi}{N+1/2}$, $j=0,\cdots,N$, be the $N+1$ points of Gauss-Radau-Chebyshev formula associated with point $x = \xi_0 = -1$. Let $P_C : C^0(\overline{I}) \rightarrow \mathbb{P}_N$ be the interpolation operator associated with these $(N+1)$ points.

Let $\tilde{u}_N(t) = P_C u(t)$, where u is the solution of problem (5.1). According to Theorem 4.1, we have $\tilde{u}_N(-1,t) = u(-1,t) = 0$ and

$$(5.8) \qquad \|u(t) - \tilde{u}_N(t)\|_{\mu,\omega_1} < C \, N^{2\mu-s} \|u(t)\|_{s,\omega},$$

where $\omega(x) \equiv (1-x^2)^{-1/2}$ is the Chebyshev weight. (Equation (5.8) follows from the fact that $\omega_1(x) < \omega(x)$, for all $x \in I$.)

Setting $z(t) = (u - \tilde{u}_N)(t)$,

$$(5.9) \qquad \frac{\partial \tilde{u}_N}{\partial t} + a(x) \frac{\partial \tilde{u}_N}{\partial x} = \frac{\partial z}{\partial t} + a(x) \frac{\partial z}{\partial x}, \qquad\qquad x \in I, \ t > 0.$$

In particular, the equation (5.9) is true for $x = x_j$, $j = 1, \cdots, N$, so setting $W_N(t) = (\tilde{u}_N - u_N)(t)$ we have

$$\left(\frac{\partial W_N}{\partial t} + a \frac{\partial W_N}{\partial x}\right)(x_j) = \left(\frac{\partial z}{\partial t} + a \frac{\partial z}{\partial x}\right)(x_j).$$

Multiplying by $\omega_j \dfrac{W_N(x_j)}{a_N(x_j)}$ and summing up from $j = 0$ to N

$$\left(W_N, \frac{\partial W_N}{\partial t}\right)_N + \int_I W_N \frac{\partial W_N}{\partial x} \omega_1 \, dx = \left(\frac{\partial z}{\partial t}, W_N\right)_N + \sum_{j=0}^{N} \omega_j \frac{\partial z}{\partial x}(x_j) W_N(x_j),$$

whence

$$\|W_N\|_N \frac{d}{dt} \|W_N(t)\|_N = \left(W_N, \frac{\partial W_N}{\partial t}\right)_N \leq \|\frac{\partial z}{\partial t}\|_N \|W_N\|_N + \|W_N\|_N \left(\sum_{j=0}^{N} \omega_j a(x_j)\left(\frac{\partial z}{\partial x}(x_j)\right)^2\right).$$

Upon simplification

$$\frac{d}{dt} \|W_N(t)\|_N \leq \|\frac{\partial z}{\partial t}\|_N + C\left(\sum_{j=0}^{N} \omega_j \left|\frac{\partial z}{\partial x}(x_j)\right|^2\right)^{1/2},$$

and using the fact that $0 < \alpha \leq a(x) \leq C$

$$\frac{d}{dt} \|W_N(t)\|_N \leq C\left(\|\frac{\partial z}{\partial t}\|_\infty + \|\frac{\partial z}{\partial x}\|_\infty\right).$$

Now, we have for $\sigma > 1/2$

$$\|v\|_{L^\infty(I)} \; < \; \|v\|_{H^\sigma(I)} \; < \; \|v\|_{\sigma,\omega} \; .$$

whence for $\sigma > \frac{1}{2}$

$$\frac{d}{dt} \|W_N(t)\|_N \; < \; C \left(\|\frac{\partial z}{\partial t}\|_{\sigma,\omega} + \|\frac{\partial z}{\partial x}\|_{\sigma,\omega} \right).$$

As $z = u - \tilde{u}_N(t)$, we have

$$\|\frac{\partial z}{\partial x}\|_{\sigma,\omega} \; < \; \|z\|_{1+\sigma,\omega} \; < \; C \, N^{2(1+\sigma)-s} \, \|u(t)\|_s \, .$$

Since

$$\frac{\partial z}{\partial t} \; = \; \frac{\partial u}{\partial t} \; - \; P_N \frac{\partial u}{\partial t} \, ,$$

we have

$$\|\frac{\partial z}{\partial t}\|_{\sigma,\omega} < C \, N^{2\sigma-(s-1)} \, \|\frac{\partial u}{\partial t}\|_{s-1,\omega} < C \, N^{2\sigma+1-s} \, \|u(t)\|_{s,\omega} \, ,$$

(where we have used equation (5.1i)).

Finally, we have

$$\frac{d}{dt} \|W_N(t)\|_N \; < \; C \, N^{2(1+\sigma)-s} \, \|u(t)\|_{s,\omega} \, ,$$

so integrating between 0 and t

$$\|W_N(t)\|_N < \|W_N(0)\|_N + C \, N^{2(1+\sigma)-s} \int_0^t \|u(t)\|_{s,\omega} \, .$$

According to (5.7), we have for $t \leqslant T$

$$\|W_N(t)\|_{\omega_1} \leqslant \frac{C}{\alpha} \|W_N(0)\|_{\omega_1} + C \, N^{2(1+\sigma)-s} .$$

Now,

(5.10) $\qquad \|W_N(0)\|_{\omega_1} = \|u_{0N}-P_C u_0\|_{\omega_1} \leqslant \|u_{0N}-u_0\|_{\omega_1} + \|u_0-P_C u_0\|_{\omega}$

$$\leqslant \|u_{0N}-u_0\|_{\omega_1} + C \, N^{-s} \|u_0\|_{s,\omega} \, ,$$

whence

$$\|W_N(t)\|_{\omega_1} \leqslant C \, \|u_{0N}-u_0\|_{\omega_1} + C \, N^{2(1+\sigma)-s} .$$

To conclude, we note that

$$\|u(t)-u_N(t)\|_{\omega_1} \leqslant \|u(t)-\tilde{u}_N(t)\|_{\omega_1} + \|W_N(t)\|_{\omega_1}$$

and that

$$\|u(t) - \tilde{u}_N(t)\|_{\omega} \leqslant C \, N^{-s} \, \|u(t)\|_{s,\omega} .$$

$\qquad\qquad\qquad\qquad\qquad\qquad\qquad\qquad\qquad\qquad\qquad\qquad$ Q.E.D.

<u>Remark 5.3</u>:

1. We may choose

$$u_{0N} = P_C u_0 .$$

In this case (see (5.10)), $W_N(0) = 0$ in the preceding discussion so that we obtain directly

$$(5.11) \qquad \| u(t) - u_N(t) \|_{\omega_1} < C \, N^{2(1+\sigma)-s}.$$

(Of course, other choices are possible.)

2. The result established in Theorem 5.2 shows the convergence of $u_N(t)$ to $u(t)$ has a known rate when $s > 3$, that is to say for very regular solutions (at least C^2); however, it might not be optimal.

Method B

We now consider collocation points

$$(5.12) \qquad x_j = -\cos \frac{j\pi}{N + \frac{1}{2}}, \qquad j = 0, \cdots, N.$$

The x_j are the points of Gauss-Radau-Chebyshev formula associated with point $x = -1$ (see Example 2.1). The numerical integration formula

$$(5.13) \qquad \int_I f(x) \, \omega(x) \, dx \simeq \sum_{j=0}^M \tilde{\omega}_j f(x_j),$$

is then <u>exact</u> for $f \in \mathbb{P}_{2N}$, with

$$\tilde{\omega}_j = \frac{2\pi}{2N+1}, \qquad j = 1, \cdots, N \quad \text{and} \quad \tilde{\omega}_0 = \frac{\pi}{2N+1}.$$

(See (2.4).)

This means that (choosing $g = (1-x)f$) the formula

(5.14) $$\int_I g(x)\omega_1(x)dx \simeq \sum_{j=0}^{N} \omega_j g(x_j),$$

where $\omega_j = (1-x_j)\tilde{\omega}_j$, is exact for $g \in \mathbb{P}_{2N-1}$.

Theorem 5.3: With the choice (5.12) for the collocation points, we have stability for the discrete norm $\|\cdot\|_N$ associated with the discrete scalar product

$$(u,v)_N = \sum_{j=0}^{N} \frac{\omega_j}{a(x_j)} u(x_j)v(x_j),$$

that is to say

$$\|u_N(t)\|_N < \|u_N(0)\|_N, \qquad \text{for all } t > 0.$$

Proof: According to (5.2) we have

$$\left(\frac{\partial u_N}{\partial t} + a \frac{\partial u_N}{\partial x}\right)(x_j) = 0, \qquad j = 1,\cdots,N.$$

Multiplying by $\omega_j \dfrac{u_N(x_j)}{a(x_j)}$ and summing for $j=1,\cdots,N$, we obtain (noting that $u_N(x_0) = 0$)

$$\sum_{j=0}^{N} \frac{\omega_j}{a(x_j)} u_N(x_j) \frac{\partial u_N}{\partial t}(x_j) + \sum_{j=0}^{N} \omega_j u_N(x_j) \frac{\partial u_N}{\partial x}(x_j) = 0,$$

i.e., (see (5.14))

$$\left(u_N, \frac{\partial u_N}{\partial t}\right)_N + \int_I u_N \frac{\partial u_N}{\partial x} \omega_1 \, dx = 0,$$

and the result follows in the same fashion as in Theorem 5.1.

Q. E. D.

We leave it to the reader to establish for these collocation points the error estimate analogous to Theorem 5.2., i.e.,

$$\| u(t) - u_N(t) \|_N < C \, N^{2(1+\sigma)-s}.$$

But here, as the formula is only exact for $g \in \mathbb{P}_{2N-1}$, we do not have the analogue of (5.7), and we cannot replace the discrete norm $\| \cdot \|_N$ by the norm $\| \cdot \|_{\omega_1}$.

6. Time Discretization Schemes

Following the analysis of Part A (Part A, Section 10) we would like for reasons of efficiency to use some <u>explicit discretization schemes</u> in time. These allow us to benefit from the Fast Fourier Transform. We consider first the general case where the collocation points x_j are arbitrary.

The Choice of a Basis for U_N

Suppose $(\psi_k)_{k=1,\cdots,N}$ is the basis in U_N defined by

(6.1) $$\psi_k(x_j) = \delta_{jk}.$$

(The ψ_k are the Lagrange polynomials.) For any $v \in U_N$, we have

$$v(x) = \sum_{k=1}^{N} z_k \psi_k(x), \qquad \text{with} \quad z_k = v(x_k).$$

Setting

$$u_N(x,t) = \sum_{k=1}^{N} y_k(t)\psi_k(x);$$

where u_N is the solution to the approximate problem (5.2), we have

$$\sum_{k=1}^{N} \frac{dy_k}{dt} \psi_k(x_j) + \sum_{k=1}^{N} a(x_j)\psi_k^-(x_j)y_k = 0,$$

i.e.,

$$\frac{dy}{dt} + Ay = 0,$$

where A is the $N \times N$ matrix the coefficients of the form

$$a(x_j)\psi_k^-(x_j), \qquad j, \ k=1,\cdots,N.$$

We wish to study some properties of the eigenvalues of matrix A. Suppose $\lambda \in Sp(A)$, we have

$$Ay = \lambda y,$$

i.e.,

$$a(x_j) \sum_{k=1}^{N} \psi_k^-(x_j)y_k = \lambda y_j = \sum_{k=1}^{N} \psi_k(x_j)y_k,$$

and setting

$$u_N = \sum_{k=1}^{N} y_k\psi_k(x) \ \in \ U_N,$$

we obtain

(6.2)
$$a(x_j) \frac{\partial u_N}{\partial x} (x_j) = \lambda \, u_N(x_j).$$

Finally, multiplying by $\dfrac{\omega_j}{a(x_j)} \overline{u}_N(x_j)$ (in general $\lambda \in \mathbb{C}$ and u_N is complex valued--recall that $\overline{u}_N$ denotes the complex conjugate of u_N) we obtain (noting that $u_N(x_0) = 0$)

$$\sum_{j=0}^{N} \omega_j \frac{\partial u_N}{\partial x} (x_j) \, \overline{u_N} (x_j) = \lambda \sum_{j=0}^{N} \frac{\omega_j}{a(x_j)} u_N(x_j) \, \overline{u_N} (x_j).$$

Now, we notice that when the x_j are defined as in method A, the left-hand side is an exact numerical integration formula so that

$$\int_I \omega_1 \frac{\partial u_N}{\partial x} \overline{u}_N \, dx = \lambda \| u_N \|_N^2 .$$

Now, using integration by parts we see that

$$2 \operatorname{Re} \int_I \frac{\partial u_N}{\partial x} \overline{u}_N \, \omega_1 \, dx = - \int_I |u_N|^2 \, \omega_1' \, dx > 0.$$

We deduce

(6.3)
$$\operatorname{Re}(\lambda) > 0.$$

In the case where the x_j are the Gauss-Radau-Chebyshev points (method B) then the numerical integration formula is also exact, so that (6.3) still holds. Furthermore, we can get an upper bound for $|\lambda|$. Let $\lambda \in \operatorname{Sp}(A)$, according to (6.2), we have

$$\sum_{j=0}^{M} \tilde{\omega}_k \frac{\partial u_N}{\partial x} (x_j) \overline{u_N} (x_j) = \lambda \sum_{j=0}^{N} \frac{\tilde{\omega}_j}{a(x_j)} u_N(x_j) \overline{u_N} (x_j),$$

where we use this time the true weight of Gauss-Radau-Chebyshev formula (see (5.13)).

We have then (using the fact that $a(x)$ is bounded)

$$\left| \int_I \frac{\partial u_N}{\partial x} u_N \omega \, dx \right| \; > \; |\lambda| \, \frac{1}{C} \int_I |u_N|^2 \, \omega \, dx$$

and so

(6.4)
$$|\lambda| \; < \; \frac{\left\| \frac{\partial u_N}{\partial x} \right\|_\omega}{\| u_N \|_\omega} \; < \; C \, N^2,$$

from the inverse inequality established in Lemma 3.2.

In practice problem (5.2) is solved using <u>explicit Runge-Kutta schemes</u> (see Part A, Section 10).

Condition (6.3) ensures the stability of the order 4 scheme for a sufficiently small time step.

The condition (6.4), obtained for the method using the Gauss-Radau-Chebyshev points, shows that it is stable for

$$\Delta t \; < \; C \, N^{-2}.$$

<u>Remark 6.1</u>: Result (6.4) is less favorable than that obtained for Fourier series (see Part A, Proposition 10.2) and leads to a more severe limitation in time step. This affects the efficiency of the method in especially if resolution requires a large N.

7. The Use of the Fast Fourier Transform

In order to use the explicit schemes advantageously (given the severe limitations on the time step due to stability) it is necessary to calculate very rapidly the product of the matrix A defined in the preceding section by a column vector y with N components.

Let us begin by considering the Gauss–Radau–Chebyshev points. Let $y = (y_k)_{k=1,\cdots,N}$ be given.

Setting

$$u_N(x) = \sum_{k=1}^{N} y_k \, \psi_k(x).$$

We have by definition

$$(Ay)_j = a(x_j) \frac{\partial u_N}{\partial x}(x_j).$$

In the first stage we use the Fast Fourier Transform to calculate the coefficients $(a_n)_{n=0,\cdots,N}$ in the expansion of u_N in Chebyshev polynomials

$$u_N(x) = \sum_{n=0}^{N} a_n \, t_n(x).$$

(This is possible because the $(x_j)_{j=0,\cdots,N}$ fixed by (5.2) are precisely the projections on the real axis of $2N+1$ roots of unity.

In the second stage, we use the formula given by Lemma 3.1 to determine the coefficient b_n in the expansion of $\frac{\partial u_N}{\partial x}$ in Chebyshev polynomials.

In the third stage we again use the Fast Fourier Transform (actually, its inverse) to calculate from b_n the values $\frac{\partial u_N}{\partial x}(x_j)$ at the collocation points x_j.

The calculation in this fashion requires $O(N \log_2 N)$ operations and multiplications (instead of $0(N^2)$ operations if we directly calculate the elements of matrix A).

Method A

We shall see that for these points we may still use the Fast Fourier Transform to evaluate Ay, for given y being known, but the argument is more subtle (see Gottlieb, [8]).

We return to the choice (5.3) of the collocation points x_j. The following result is fundamental.

Proposition 7.1: <u>Let</u> $(x_j)_{j=0,\cdots,N+1}$ <u>be given by</u>

$$x_j = \cos \frac{j\pi}{N+1} .$$

<u>Suppose</u> $u \in \mathbb{R}_N$ <u>is given.</u> <u>We have</u>

$$u(x) = \sum_{n=0}^{N} a_n t_n(x),$$

<u>with</u>

$$a_n = d_n + \frac{2}{\gamma_n} (-1)^{N+n} d_{N+1}, \qquad n=0,\cdots,N,$$

<u>where the</u> $(d_n)_{n=0,\cdots,N+1}$ <u>are the expansion coefficients in terms of Chebyshev polynomials of</u> $v \in \mathbb{R}_{N+1}$ <u>such that</u>

$$v(x_j) = u(x_j), \qquad\qquad j=0,\cdots,N-1$$

(7.1)

$$v(x_{N+1}) = 0$$

and

$$\gamma_n \equiv \begin{cases} 2 & \text{for } n = 0 \text{ or } N+1 \\ 1 & \text{for } 1 \leqslant n \leqslant N. \end{cases}$$

Before proving this result, let us first explain how we use it. As the polynomial $u \in \mathbb{P}_N$ is determined by its values at the points x_j, $j=0,\cdots,n$ we cannot directly use the Fast Fourier Transform to calculate the a_n. (In fact, the $(x_j)_{j=0,\ldots,N}$ constitute only $N+1$ of the needed $N+2$ projections on the real axis of the roots of unity of order $2N+2$).

To circumvent this difficulty, we will calculate the coefficients of a polynomial $v \in \mathbb{P}_{N+1}$ which coincides with u at $(x_j)_{j=0,\ldots,N}$ and which equals to 0 at $N+2^{nd}$ point x_{N+1}. Thus, the coefficients d_n of the expansion of v in Chebyshev polynomials will be calculable using the Fast Fourier Transform.

Now, it turns out that there is a simple relation (given by Proposition 7.1) between the a_k and the d_k.

In order to prove Proposition 7.1, we need the two following results:

Lemma 7.1: Let ω be such that $\omega^{2N+2} = 1$, then

$$\frac{1}{2N+2} \sum_{n=-N}^{N+1} \omega^n = \begin{cases} 1 & \text{if } \omega = 1 \\ 0 & \text{otherwise} \end{cases} \quad .$$

(The verification of this lemma is left to the reader.)

Lemma 7.2:   <u>We have</u>

$$\sum_{n=0}^{N+1} \frac{1}{\gamma_n} (-1)^n t_n(x_j) = 0, \qquad \underline{\text{for}} \quad j=0,\cdots,N,$$

(<u>where the</u> γ_n <u>are defined in Proposition 7.1.</u>)

<u>Proof:</u>   According to Lemma 7.1 (applied with $\omega = e^{ik\frac{\pi}{N+1}}$), we have

$$\sum_{n=-N}^{N+1} e^{ink\frac{\pi}{N+1}} = 0, \qquad \text{for} \quad 1 < k \leqslant 2N+1.$$

Let us set $k = N+j+1$, with $j=0,\cdots,N$; we have

$$e^{ink\frac{\pi}{N+1}} = e^{in\frac{N+j+1}{N+1}\pi} = e^{in\pi} e^{in\frac{j\pi}{N+1}} = (-1)^n e^{in\frac{j\pi}{N+1}},$$

whence

$$\sum_{n=-N}^{N+1} (-1)^n e^{in\frac{j\pi}{N+1}} = 0.$$

Taking the real part of this relation, we obtain

$$0 = \sum_{n=-N}^{N+1} (-1)^n \cos\frac{nj\pi}{N+1} = 2 \sum_{n=0}^{N+1} \frac{(-1)^n}{\gamma_n} \cos nj \frac{\pi}{N+1} = 2 \sum_{n=0}^{N+1} \frac{(-1)^n}{\gamma_n} t_n(x_j),$$

which is the desired result.

Q. E. D.

<u>Proof of Proposition 7.1:</u>   Let $v \in P_{N+1}$ satisfy (7.1) and $(d_n)_{n=0,\cdots,N+1}$ be its Chebyshev coefficients. From Lemma 7.2, we have

$$v(x_j) = \sum_{n=0}^{N} d_n\ t_n(x_j) + d_{N+1}\ t_{N+1}(x_j)$$

$$+ (-1)^N\ 2d_{N+1}\Big[\sum_{n=0}^{N+1}\ \frac{1}{\gamma_n}\ (-1)^n\ t_n(x_j)\Big],$$

for $j=0,\cdots,N$, i.e.,

$$v(x_j) = \sum_{n=0}^{N} d_n\ t_n(x_j) + (-1)^N\ 2d_{N+1} \sum_{n=0}^{N}\ \frac{1}{\gamma_n}\ (-1)^n\ t_n(x_j).$$

The right-hand side is a polynomial of degree N which coincides with u at the $(N+1)$ points $(x_j)_{j=0,\cdots,N}$; we have then

$$a_n = d_n + \frac{2}{\gamma_n}\ (-1)^{N+n}\ d_{N+1}\ .$$

Q. E. D.

8. Solutions of the Heat Equation

We consider the equation

$$i)\quad \frac{\partial u}{\partial t} - a(x)\ \frac{\partial^2 u}{\partial x^2} = 0 \qquad\qquad x \in I,\ t > 0,$$

(8.1) $\qquad ii)\quad u(-1,t) = g(t),\ u(1,t) = h(t),\quad t > 0$

$$iii)\quad u(x,0) = u_0(x), \qquad\qquad x \in I.$$

The boundary conditions (8.1ii) are not periodic, unlike the problem considered in Part A.

We consider for simplicity the case of homogeneous boundary conditions, $g(t) = h(t) = 0$ [1] to approximate problem (8.1) with the following collocation method.

Suppose V_N is the space (of dimension $N-1$) defined by

$$V_N = \{p \; \varepsilon \; P_N : p(-1) = p(1) = 0\}$$

and $(x_j)_{0 < j < N}$ the $(N+1)$ points defined by

(8.2) $$x_j = \cos \frac{j\pi}{N} \; , \qquad\qquad j=1,\cdots,N-1;$$

we define the approximate problem by

Find $u_N(t) \; \varepsilon \; V_N$ such that

i) $$\left(\frac{\partial u_N}{\partial t} - a \frac{\partial^2 u_N}{\partial x^2}\right) (x_j) = 0, \qquad 1 < j < N-1$$

(8.3)

ii) $$u_N(x_j,0) = u_0(x_j), \qquad 1 < j < N-1.$$

We establish first the stability of the method (see Gottlieb [8]).

(In the present section, all functions are supposed real-valued, and coefficient $a(x)$ is supposed regular and strictly positive).

Theorem 8.1: Let $(\omega_j)_{0 < j < N}$ denote the coefficients of the $(N+1)$-points Gauss-Lobatto-Chebyshev formula and $\|\cdot\|_N$ the discrete norm defined by

---

(1) Note that in this particular case the Fourier method is applicable.

$$\| p \|_N \equiv \left((p,p)_N \right)^{1/2}$$

$$(p,q)_N \equiv \sum_{j=0}^{N} \frac{\omega_j}{a(x_j)} \; p(x_j) \overline{q(x_j)},$$

then we have

$$\| u_N(t) \|_N \leq \| u_N(0) \|_N.$$

We will need the following lemma from Gottlieb and Orszag [9]):

Lemma 8.1:   <u>Let</u> $u \in C^1(\overline{I})$ <u>be a function such that</u>

$$u(1) = u(-1) = 0$$

<u>then, we have</u>

$$\int_I u \; u'' \; \omega \; dx \leq 0,$$

<u>where</u> $\omega(x) \equiv (1-x^2)^{-1/2}$ <u>denotes the Chebyshev weight.</u>

<u>Proof of Lemma 8.1:</u>   We note first that as u is Lipschitz continuous and zero at the end points, we have

$$\lim_{x \to \pm 1} \omega(x)u(x) = 0.$$

It follows that

$$\int_I u \; u'' \; \omega \; dx = -\int_I (\omega u)' u' dx.$$

Now,

$$\int_I (\omega u)^\sim u^\sim dx = \int_I (\omega u)^\sim (\omega u)^\sim \frac{1}{\omega} dx - \int_I (\omega u)^\sim \frac{\omega^\sim}{\omega} u \, dx,$$

and

$$\int_I (\omega u)^\sim \frac{\omega^\sim}{\omega} u \, dx = \int_I (\omega u)^\sim \omega u \frac{\omega^\sim}{\omega^2} dx$$

$$= \frac{1}{2} \int_I (\omega^2 u^2)^\sim \frac{\omega^\sim}{\omega^2} dx = -\frac{1}{2} \int_I \omega^2 u^2 \left(\frac{\omega^\sim}{\omega}\right)^\sim dx,$$

where we have integrated by parts, to obtain the last term. Now,

$$\left(\frac{\omega^\sim}{\omega}\right)^\sim = \left(x(1-x^2)^{-1/2}\right)^\sim = (1-x^2)^{-3/2},$$

yields

$$\int_I (\omega u)^\sim \frac{\omega^\sim}{\omega} u \, dx = -\frac{1}{2} \int_I (1-x^2)^{-5/2} u^2 dx.$$

We have shown that

$$\int_I u \, u'' \, \omega \, dx = -\int_I \left((\omega u)^\sim\right)^2 \frac{1}{\omega} dx - \frac{1}{2} \int_I (1-x^2)^{-5/2} u^2 \, dx,$$

and the result follows.

Q. E. D.

Proof of Theorem 8.1: According to (8.3i) and the definition of V_N, we have

$$\sum_{j=0}^N \frac{\omega_j}{a(x_j)} \frac{\partial u_N}{\partial t}(x_j) u_N(x_j) - \sum_{j=0}^N \omega_j \frac{\partial^2 u_N}{\partial x^2}(x_j) u_N(x_j) = 0,$$

so using the fact that the $(N+1)$-point Gauss-Lobatto-Chebyshev formula is exact for the polynomials of degree $< 2N$

$$(\frac{\partial u_N}{\partial t}, u_N)_N - \int_I \frac{\partial^2 u_N}{\partial x^2} u_N \omega \, dx = 0.$$

With Lemma 8.1, we obtain

$$\frac{d}{dt} \| u_N(t) \|_N^2 = (\frac{\partial u_N}{\partial t}, u_N)_N < 0,$$

and hence the result.

Q. E. D.

An Error Estimate

Theorem 8.2: Let $\sigma > \frac{1}{2}$, $s > 2\sigma + 4$ and ϵ such that $0 < \epsilon < T$ be given; then if $u \in L^1(0,T;H^s_\omega(I))$, there exists a constant C such that

$$\| u(t) - u_N(t) \|_N < C N^{2\sigma+4-s}, \qquad \epsilon < t < T.$$

Proof: Let $\tilde{u}_N(x,t) = (\pi_C u)(x,t)$, where π_C is the interpolation operator defined in Section 4.

Let $z = \tilde{u}_N - u$. We have, using equation (8.11),

$$\frac{\partial}{\partial t} \tilde{u}_N - a(x) \frac{\partial^2}{\partial x^2} \tilde{u}_N = \frac{\partial}{\partial t} z - a(x) \frac{\partial^2 z}{\partial x^2}.$$

Let then $W_N(t) = (\tilde{u}_N - u_N)(t) \in V_N$; we have, with (8.1i),

$$\left(\frac{\partial W_N}{\partial t} - a \frac{\partial^2}{\partial x^2} W_N\right)(x_j) = \left(\frac{\partial z}{\partial t} - a \frac{\partial^2 z}{\partial x^2}\right)(x_j) , \qquad 1 \leq j \leq N-1.$$

Multiplying the two sides by $\frac{\omega_j}{a(x_j)} W_N(x_j)$ and summing from $j=1$ to $N-1$, we obtain

$$\left(\frac{\partial W_N}{\partial t}, W_N\right)_N + \int_I \frac{\partial^2 W_N}{\partial x^2} W_N \, \omega \, dx = \left(\frac{\partial z}{\partial t} - a \frac{\partial^2 z}{\partial x^2}, W_N\right)_N,$$

whence, with Lemma 8.1

$$\frac{1}{2} \|W_N\|_N \frac{d}{dt} \|W_N(t)\|_N = \left(\frac{\partial W_N}{\partial t}, W_N\right)_N \leq \left(\frac{\partial z}{\partial t} - a \frac{\partial^2 z}{\partial x^2}, W_N\right)_N$$

$$\leq \|\frac{\partial z}{\partial t} - a \frac{\partial^2 z}{\partial x^2}\|_N \|W_N\|_N,$$

and by integration from 0 to t (and using the fact that $\|W_N(0)\|_N = 0$)

$$\|W_N(t)\|_N \leq \int_0^t \|\left(\frac{\partial z}{\partial t} - a \frac{\partial^2 z}{\partial x^2}\right)(\tau)\|_N d\tau .$$

Now,

$$\|\frac{\partial z}{\partial t} - a \frac{\partial^2 z}{\partial x^2}\|_N \leq C\|\frac{\partial z}{\partial t} - a \frac{\partial^2 z}{\partial x^2}\|_{\sigma,\omega} ,$$

since $H_\omega^\sigma(I) \to H^\sigma(I) \to L^\infty(I)$ for $\sigma > 1/2$. From Theorem 4.2, we have

$$\|a \frac{\partial^2 z}{\partial x^2}\|_{\sigma,\omega} \leq C\|u - \tilde{u}_N\|_{\sigma+2,\omega} \leq C \, N^{2(\sigma+2)-s} \|u\|_{s,\omega}$$

and

$$\|\frac{\partial z}{\partial t}\|_{\sigma,\omega} = \|\frac{\partial u}{\partial t} - \pi_C \frac{\partial u}{\partial t}\|_{\sigma,\omega} \leq C N^{2\sigma-(s-2)} \|\frac{\partial u}{\partial t}\|_{s-2,\omega}.$$

Now, from equation (8.1i), we have

$$\|\frac{\partial u}{\partial t}\|_{s-2,\omega} = \|a \frac{\partial^2 u}{\partial x^2}\|_{s-2,\omega} \leq C\|u\|_{s,\omega}.$$

Finally, we have shown that

$$\|(\frac{\partial z}{\partial t} - a \frac{\partial^2 z}{\partial x^2})(\tau)\|_{\sigma,\omega} \leq C N^{2\sigma+4-s} \|u(\tau)\|_{s,\omega},$$

so for $0 < t < T$

$$\|(u - u_N)(t)\|_N \leq \|(u - \tilde{u}_N)(t)\|_N + \|W_N(t)\|_N$$

$$\leq C\|(u-\tilde{u}_N)(t)\|_{\sigma,\omega} + C N^{2\sigma+4-s} \int_0^t \|u(\tau)\|_{s,\omega} d\tau.$$

Now, for $t \geq \epsilon > 0$, we know that $u(t) \in H_\omega^s(I)$ for all s (regularizing effect), therefore

$$\|(u-\tilde{u}_N)(t)\|_{\sigma,\omega} \leq C N^{2\sigma-s}.$$

Finally, as we have assumed that $u \in L^1(0,T; H_\omega^s(I))$, we obtain

$$\|(u - u_N)(t)\|_N \leq C(N^{2\sigma-s} + N^{2\sigma+4-s}),$$

which yields the desired result.

Q. E. D.

Remark 8.1: In order that the solution u of the heat equation (8.1) belong to $L^1(0,T; H_\omega^s(I))$, it is necessary that the initial solution should satisfy certain regularity and compatibility conditions and also the boundary conditions (see Bramble-Schatz-Thomee [2]).

References

[1] Auslander-Tolimieri: "Is Fast Fourier Transform pure or applied mathematics," Bull. (New Series) AMS, 1, 6 (1979), pp. 847-898.

[2] Bramble-Schatz- Thomee: SIAM J. Numer. Anal., 14 (1977), pp. 218-241.

[3] Brutman, L.: "On the Lebesgue function for polynomial interpolation," SIAM J. Numer. Anal. 15 (1978), pp. 694-704.

[4] Canuto, C. and A. Quarteroni: "Proprietés d´approximation dans les espaces de Sobolev de systemes de polynômes orthogonaux," C.R. Acad. Sci. Paris 290 (1980) Series A, pp. 925-928, see also "Approximation results for orthogonal polynomials" in Math. Comp. 38 (1982), pp. 67-86.

[5] Canuto, C., M. Y. Hussaini, A. Quarteroni, and T. A. Zang: Spectral Methods in Fluid Dynamics, Springer-Verlag, 1987.

[6] Carleson: Acta Mathematica 116 (1966), pp. 135-157.

[7] Davis, P. J. and P. Rabinowitz: Methods of Numerical Integration, Academic Press, New York, 1975.

[8] Gottlieb, D.: "The stability of pseudospectral Chebyshev methods," Math. Comp. 36 (1981), pp. 107-118.

[9]   Gottlieb, D. and S. A. Orszag: <u>Numerical Analysis of Spectral Methods</u>,
       SIAM, Philadelphia, 1977.

[10]   Kato, T.: <u>Perturbation Theory of Linear Operators</u>, Springer-Verlag,
       Berlin, 1980.

[11]   Lions, J. L. and E. Magenes: <u>Nonhomogeneous Boundary Value Problems</u>,
       Springer-Verlag, Berlin, 1972.

[12]   Majda, A., J. McDonough, and S. Osher: "The Fourier method for
       nonsmooth initial data," Math. Comp. **32** (1978), pp. 1041-1081.

[13]   Pasciak, J.: "Spectral and pseudospectral methods for advection
       equations," Math. Comp. **35**, 152 (1980), pp. 1081-1092.

[14]   Richtmyer, R. and K. Morton: <u>Difference Methods for Initial Value
       Problems</u>, J. Wiley & Sons, New York, 1963.

[15]   Rivlin, T. J.: <u>An Introduction to the Approximation of Functions</u>,
       Dover, New York, 1969.

[16]   Schwartz, L.: <u>Theorie des Distributions</u>, Hermann, Paris, 1967.

[17]   Taylor, M.: <u>Pseudo Differential Operators</u>, Springer, Lecture Notes in
       Mathematics No. 416.

[18]   Trèves, F.: <u>Topological Vector Spaces, Distributions, and Kernels</u>,
       Academic Press, New York, 1967.